John James Wild

Thalassa

An essay on the depth, temperature, and currents of the ocean

John James Wild

Thalassa

An essay on the depth, temperature, and currents of the ocean

ISBN/EAN: 9783337037130

Printed in Europe, USA, Canada, Australia, Japan

Cover: Foto ©berggeist007 / pixelio.de

More available books at **www.hansebooks.com**

AN ESSAY

ON THE

Depth, Temperature, and Currents of the Ocean

BY

JOHN JAMES WILD

Member of the Civilian Scientific Staff of H.M.S. "Challenger"

With Charts and Diagrams by the Author

London:

MARCUS WARD & CO., 67 & 68, CHANDOS STREET, STRAND
AND ROYAL ULSTER WORKS, BELFAST

1877

CONTENTS.

PAGE

PREFACE, ... 7

CHAPTER I.

DEPTH OF THE OCEAN.

Distribution of Land and Water, 11—Principal Areas of Elevation and of Depression, 12—Results of Soundings obtained to present Date, 13—Average Depth of the Ocean, 14—Greatest Depths ascertained in the Atlantic, 14; in the Indian Ocean and in the Pacific, 15—Configuration of the Sea-bottom, 16—The Basin of the Atlantic, 17—Central Atlantic Plateau, 18—The Basin of the Indian Ocean, 19—The Kerguelen Plateau, 19—The Basin of the Pacific Ocean, 20—Submerged Plateau in the South-Eastern Pacific, 21—Large Area of Depression upon the Limits of the South Pacific and the Southern Ocean, 22—The Basin of the Southern Ocean, 23—The Basin of the Antarctic Sea, 24—The Basin of the Arctic Sea, 25—Area of Depression between Greenland and Norway, 26.

CHAPTER II.

TEMPERATURE OF THE OCEAN.

Surface Temperature, 27—Total Range of Oceanic Temperatures, 28—Influence of Currents upon the Distribution of Temperature, 29—Oceanic Currents an Adequate Cause of Alterations of Climate, 29—Deep-Sea Temperature, 30—Earliest Systematic Observation of Deep-Sea Temperature, 30—Effect of Pressure upon the ordinary Thermometer, 31—The Miller-Casella Thermometer, 32—Serial Soundings and Temperature Curves, 34—Construction of Temperature Curves, 35—Deductions from the Curve, 36—General Decrease of Temperature from the Surface to the Bottom, 37—

Distribution of Temperature in nearly land-locked Basins, 38—Increase of Temperature from the Surface towards the lower Strata in the Polar Regions, 39—Distribution of Oceanic Temperature effected through the Agency of Currents, 40—Formation of Intermediate Strata, 41—Connection between the Gradients of the Curve and the relative Motion of the different Strata, 42—Illustrations from the Temperature Soundings of H.M.S. "Challenger," 42-45.

CHAPTER III.

CURRENTS OF THE OCEAN.

The Aqueous and the Aerial Oceans, 46—Resemblance between the Phenomena in the two Terrestrial Envelopes, 47—Thermal Circulation, 47—Disproportion between the Vertical and the Horizontal Extension of the Aqueous and Aerial Envelopes, 48—Conclusions based upon this Disproportion, 49—Influence of the several Areas of the Earth's Surface upon Oceanic and Atmospheric Circulation, 50—Parallelism between Oceanic and Atmospheric Currents, 50—Formation of Belts of Calms and Belts of Currents, 51—The Critical Latitudes, 52—Effect of the present Distribution of Land and Water upon Atmospheric and Oceanic Circulation, 52—Subdivision of the latter into Distinct Areas of Circulation, 52—Atmospheric and Oceanic Currents revolve round Areas of High Barometric Pressure, 53—Effect of the Sun's apparent Progress from Tropic to Tropic upon Atmospheric and Oceanic Currents, 53—Surface and Under-Currents, 54—Relation between the Colour of Sea-water and the Percentage of Salt held in solution, 54—Limits of the Specific Gravity of Salt Water according to Temperature and to Percentage of Salt, 55—Change of a Surface-Current into an Under-Current, or Vertical Circulation of the Oceanic Waters, 56— Diagram of the System of Circulation in an Oceanic Basin, 57.

CHAPTER IV.

THE TEMPERATURE SECTIONS SURVEYED BY H.M.S. "CHALLENGER" IN THE ATLANTIC.

Explanation of Diagrams and Tables, 58—Section from Teneriffe to Sombrero, 59—Contrast between the Eastern and Western Basin of the North Atlantic, 61—Observed Rise of the Isotherms with the Sea-bottom, 62—Section from St. Thomas to Halifax, 63—The Gulf Stream a branch of the North Atlantic Equatorial Current, 65—Parallelism between the Gulf Stream and the Kuro-Siwo Current, 66—Area of Alternate Streaks of Warm and Cold Water between Halifax and Bermudas, 67—Section between Cape May and Madeira, 69—Temperature of the Gulf Stream, 71—Cold Current off the Azores, 72—The Sargasso Sea, 73—Section from Madeira to Tristan d'Acunha, 74—The Equatorial Belt, 75—Cold Under-Current across the Equator, 77—Isothermal and Isobathymetrical Lines, 79—Section from Cape Palmas to Cape S. Roque, 80—Contrast between the Eastern and Western Basin of the South Atlantic, 81—Section between Cape S. Roque and Tristan d'Acunha, 81—Section from the Falkland Islands to the Cape of Good Hope, 83—The Antarctic Under-Current of the South Atlantic, 85—The Equatorial Current of the South Atlantic, 86—The Antarctic Surface-Current, 86.

CHAPTER V.

THE TEMPERATURE SECTIONS SURVEYED BY H.M.S. "CHALLENGER" IN THE SOUTHERN OCEAN, THE INDIAN ARCHIPELAGO, AND THE PACIFIC.

Section from the Cape of Good Hope to the Ice-barrier and to Melbourne, 88—The Agulhas Current, 89—Sudden Changes of Temperature in Simons Bay and Table Bay, 91—Encounter of Warm and Cold Currents off the Cape of Good Hope, 93—Icebergs in the Southern Ocean, 95—The South Australian Current, 96—Section from Sydney to Cook Strait, New Zealand, and from Cook Strait to Tonga Tabu, 97—The Basin between Australia and New Zealand, 99—Sections from Tonga Tabu to Torres Strait, 100—The Melanesian Sea, 101—Conclusions derived from the Temperature Conditions of the latter, 103—Sections from Torres Strait to Hong-kong, and from Hong-kong to the Admiralty Islands, 104—The Arafura Sea, 104—Existence of a Deep Channel between the Papua-Australian Plateau and the Plateau of the Indian Archipelago, 105—The Banda Sea, 105—The Molucca Passage, 107—The Sea of Celebes, 107—The Sulu Sea, 107—The Philippine Inland Seas, 108—The China Sea, 108—The Sea of Papua, 108—Section from the Admiralty Islands to Japan, 109—The Sea of Magallanes, 109—The Kuro-Siwo Current and the Gulf Stream compared, 111—The Arctic Current of the North Pacific, 113—Area of Alternate Streaks of Warm and Cold Water south of Japan, 115—Section from Yokohama to Station 253, 115—The Kuro-Siwo Current and the Arctic Current, 116—Area of Alternate Streaks of Warm and Cold Water east of Japan, 116—Section from Station 253, along the Meridian of Honolulu and Tahiti, to Station 288, 117—The Pacific and the Atlantic Oceans compared, 117—The Equatorial Counter Currents of the Atlantic, the Indian, and the Pacific Oceans, 119—Section from Station 288 to Valparaiso and Magellan Straits, 121.

CHAPTER VI.

THE BED OF THE OCEAN.

Changes in the Distribution of Land and Water, 124—Average Height of the Dry Land, 126—Low Angle of Inclination of the Sea-bottom, 127—Formation of Sub-oceanic Strata, 128—Marginal and Central Deposits, 129—Formation of Central Oceanic Plateaux, 130—Deposits composed of Inorganic and Organic Particles, 131—Absence of Organic Remains no Evidence of the Antiquity of a Geological Formation, 133—Formation of Areas of Elevation and of Areas of Depression, 134—Formation and Transformation of Continents, 135—Primary and Secondary Areas of Elevation, 136—Formation of Mountain Ranges and Submarine Ridges, 137—Oceanic Pressure a Cause of Internal Heat, 138—Elevation of Mountain Ranges and Creation of Axis of Volcanic Eruption, due to Lateral Pressure, 139—Water the Principal Agent in the Disintegration, the Re-distribution, and the Accumulation of the Solid Matter composing the Earth-Crust, 140.

List of Illustrations.

CHARTS.

	PAGE
PLATE 1.—The Northern and Southern Hemispheres, with Track of H.M.S. "Challenger,"	11
PLATE 2.—Contour-Chart of the Bottom of the Ocean,	14
PLATE 3.—Contour-Chart of the Bottom of the Atlantic,	17
PLATE 4.—Surface-Currents and Surface-Temperatures,	27
PLATE 4A.—Lines of Equal Barometric Pressure (Isobars) for July, August, September,	46
PLATE 5.—Current Chart of the Ocean,	53

OCEANIC SECTIONS.

	PAGE
Approximate Section through the Atlantic,	*Frontispiece.*
PLATE 6.—From Teneriffe to Sombrero,	60
PLATE 7.—From St. Thomas to Halifax,	64
PLATE 8.—From Cape May to Madeira,	70
PLATE 9.—From Madeira to Tristan d'Acunha,	76
PLATE 10.—From Cape Palmas to Cape S. Roque and Tristan d'Acunha,	82
PLATE 11.—From the Falkland Islands to the Cape of Good Hope,	84
PLATE 12.—From the Cape of Good Hope to Cape Otway,	90
PLATE 13.—From Lat. 50° S. to the Antarctic Circle,	96
PLATE 14.—From Port Jackson to Cook Strait and the Friendly Islands,	98
PLATE 15.—From the Fiji Islands to Torres Strait,	102
PLATE 16.—From Torres Strait to Hong-kong and to the Admiralty Islands,	106
PLATE 17.—From the Admiralty Islands to Japan,	110
PLATE 18.—From Yokohama to Station 253,	114
PLATE 19.—From Station 253 to Station 288,	118
PLATE 20.—From Station 288 to the Coast of Chile,	122

DIAGRAMS.

	PAGE
✓The Miller-Casella Thermometer, ...	32
Fig. 1.—Temperatures in the South Atlantic, Fig. 2.—Temperatures in the Gulf Stream and Arctic Current,	41
Fig. 3.— Fig. 4.— } Temperatures in the North Atlantic,	42
Fig. 5.—Temperatures in the Agulhas Current, Fig. 6.—Temperatures in the South Australian Current,	43
Fig. 7.—Temperatures in the Antarctic Current, Fig. 8.—Temperatures in the Southern Ocean,	44
Fig. 9.—Atlantic Equatorial Temperatures, Fig. 10.—Pacific Equatorial Temperatures,	45
✓Fig. 11.—Diagram of Oceanic Isotherms, ✓Fig. 12.—Diagram of Oceanic Circulation,	57
Fig. 13.—Diameter of Rotation, Fig. 14.—Gradients of the Sea-bottom,	124

PREFACE.

THE numerous and successful efforts made in recent times to extend, and as far as possible to complete, our knowledge of the globe we inhabit, constitute one of the most characteristic features of the present age. The central parts of great continents, hitherto untrodden by the foot of civilised man, are only now commencing to be systematically explored, and, while the interest of the general reader is stimulated from time to time by the description of newly-discovered regions, a rich harvest of fresh materials is placed at the disposal of the scientific student.

The work carried on with so much energy and success on *terra firma* has been supplemented in the domain of the sea by several naval expeditions, sent out for the especial purpose of fathoming the depths of the ocean, of observing the currents and the physical and chemical condition of the water, of bringing up from the bottom samples of the deposits now in process of formation, and of gathering specimens of the countless forms of animal life with which the sea abounds at all depths.

As a natural consequence of the new facts brought to light day after day, opinions held until recently by the most competent authorities in almost every branch of Natural Science, but especially in Biology, Geology, and Physical Geography, have undergone considerable modifications, or have had to be abandoned altogether. On the other hand, the numerous carefully-made observations collected in every quarter of the globe furnish an opportunity for attempting, with renewed chances of success, the solution of several problems, for which no complete answer has as yet been found. One of these problems is that of " Oceanic Currents." Ever since Lieutenant Maury, in his admirable work on the Physical Geography of the Sea, undertook to explain the probable causes of the "beautiful system of circulation" by which "cooling streams are brought from the Polar Seas to temper the heat of the Torrid Zone," this question has been the subject of frequent controversy; but I believe I am correct in stating that none of the theories advanced in explanation of this gigantic natural phenomenon have met with general acceptance, nor has any satisfactory solution of the problem been offered.

One of the principal causes of this failure will be found in the want which existed, until within late years, of the necessary appliances and opportunities for ascertaining the conditions which prevail in every part of the ocean, from its surface down to the bottom. The scientific explorer possessed neither proper sounding apparatus, nor instruments capable of resisting the enormous pressures to which they are subject at great depths, nor the aid of steam-power for expediting the hauling in of miles of sounding-line or dredge-rope; and last, not least, 'he found no Government willing to open the national purse in favour of scientific experiments planned on a scale the expense of which exceeded the resources of even wealthy individuals, and which were absolutely beyond reach of the proverbially poor devotee of Science. In the absence of these indispensable helps, it was next to impossible to secure observations sufficiently reliable to afford a criterion in judging between the numerous theories which have been advanced, however sound the premises and logical the deductions on which they were based.

Another source of failure may be traced to the attempt to explain a highly-complex phenomenon by the operation of a single cause, and to the consequent neglect of one or other of the numerous conditions which must more or less determine the direction, velocity, volume, and persistency of oceanic currents—such, for example, as the unequal distribution of solar heat over the surface of our planet, the agency of the winds or atmospheric currents, the difference of temperature, specific gravity and chemical composition of the water, the direction of the coast-lines, the distribution of land and water in general, the configuration of the sea-bottom, &c. It remains to the student to ascertain, with the help of the observations at his disposal, the part which is to be assigned to each of these conditions in the great phenomenon of oceanic circulation, considered as the final product of various causes, all acting and reacting upon each other.

Amongst the principal efforts in the domain of deep sea exploration, we have the labours of the officers of the United States Coast Survey along the course of the Gulf Stream (1845 to 1859), the soundings of the U.S.S. "Mercury" between Barbadoes and Sierra Leone in 1871, the observations made on board H.M.S. "Lightning" and H.M.S. "Porcupine" in the seas extending from the Færoe Islands to the Mediterranean (1868-1870), culminating finally in the two voyages of circum-navigation made by H.M.S. "Challenger" (1872-1876) and the German frigate "Gazelle" (1874-1876). The extensive series of soundings for which we are indebted to these expeditions has received further additions through the operations of H.M.S. "Valorous" between the English Channel and Davis Strait in 1875, of the Norwegian ship "Vöringen" between Norway and Iceland, of the U.S.S. "Gettysburg" in the North Atlantic in 1876, the U.S.S. "Tuscarora" in the Pacific Ocean (1874 to 1876), and the latest Arctic Expedition under the command of Sir G. S. Nares (1875-1876).

In the charts and diagrams which accompany the following pages, I have endeavoured to combine the results of recent observations, and more especially of the sounding operations carried on on board H.M.S. "Challenger" during her cruise round the world, in so far as they throw any light upon the distribution of depth, temperature, and currents in the different oceanic basins which have been explored.

The advance made in our days towards a satisfactory solution of the problem of oceanic circulation will probably be recorded with due completeness by abler hands; meantime, I am not without hope that the contents of this essay may find a welcome amongst those who have followed with interest the progress of the numerous expeditions sent out of late years to clear up the mysteries of the ocean.

JOHN JAMES WILD.

LONDON, *May*, 1877.

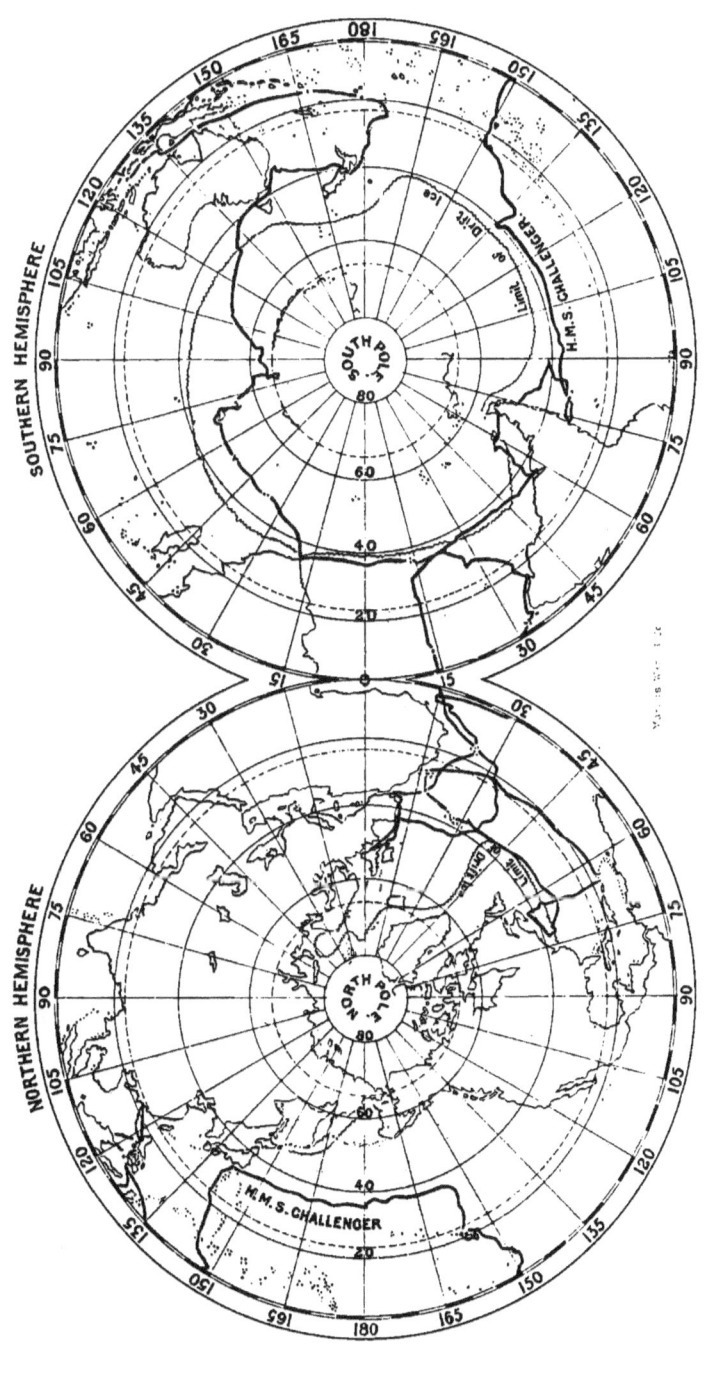

THALASSA.

CHAPTER I.

DEPTH OF THE OCEAN.

Distribution of Land and Water—Depth of the Ocean—Configuration of the Sea-bottom—Description of the Basin of the Atlantic—The Indian Ocean—The Pacific—The Southern Ocean—The Arctic Sea—The Antarctic Sea.

DISTRIBUTION OF LAND AND WATER.—Our conception of the relative distribution of land and water over the surface of the Earth has been hitherto limited to a comparison of the superficial areas occupied by these two elements, such as they are presented to us on a chart of the world. In this sense we speak of the different continents and islands which constitute the sum total of dry land, and of the different oceans and seas which compose the water-surface of our planet. But if we wish to form a more perfect idea of the distribution of land and water, we must consider not only the length and breadth of the areas occupied, but also the height of the land and the depth of the water; in other words, the volume of those portions of the solid crust of the earth which are raised above the level of the sea, and the volume of the masses of water which fill up the depressed portions of the earth's crust. We are thus led to regard the surface of the solid crust of our planet as composed of heights and hollows, of areas of elevation and areas of depression, and, as a next step, to discriminate between these areas—not according to the usual standard of the level of the sea, but according to their relative distance from the centre of the earth. In this sense we may conceive an area of elevation—*i.e.*, a raised portion of the earth's surface, which may be partially or entirely covered with water, and an area of depression—*i.e.*, a hollow in the same surface, which may be raised above the level of the sea, and form dry land or the basin of an inland sea or lake.

If we examine the chart of the world (Plates 1 and 2) in the light which has been thrown upon this question by all the reliable soundings obtained up to the present, it will be found that continents and islands which we have been in the habit of considering as separated from each other by wide seas and deep straits virtually form part of the same area of elevation; and, in a similar manner, that certain oceans and seas, which we are accustomed to distinguish by separate names, form part of the same area of depression. It will also appear that, with the exception of the islands scattered over the face of the ocean and of the Antarctic region, all the dry land at present existing may be reduced to one large area of elevation, gravitating towards the North Pole, as the common centre of the principal land masses; similarly, if we except the Arctic region and other inland basins, all the oceans and seas compose a single vast area of depression with the South Pole for common centre of the larger accumulations of water on this globe. The Arctic region forms a distinct area of depression placed in the centre of the great area of elevation, and the Antarctic region, according to the evidence we at present possess, is an area of elevation, surrounded on all sides by the above-described great area of depression. The numerous small islands that crop up in the middle of the oceanic basins are generally found associated in groups, and they belong to areas of elevation at the present time submerged, that is to say, in the condition in which we know the dry land to have been at an epoch more or less remote in the history of our planet.

Distribution of Land and Water.

In support of the above generalisation, we may point to the following facts as established by recent soundings (Plates 2 and 3). The 100-fathom line, as is well known, joins the whole of the British Islands, including the Hebrides, Orkneys, and Shetland Islands, to the continent of Europe. It forms a broad band connecting the Asiatic and American continents across Behring Strait. It unites Australia, Papua, and Tasmania in a single area of elevation, which, together with the intervening archipelago of Java, Sumatra, Borneo, Celebes, the Moluccas, and the Philippines, may be looked upon as a prolongation of the continent of Asia. It joins Ceylon to Hindostan, and the Falkland Islands to the South American continent. The 500-fathom line connects North America, Greenland, Iceland, the Færoe Islands, and the continent of Europe, the only unexplored space being Denmark Strait, between Iceland and Greenland, where the soundings may exceed the above depth. The 1000-fathom line unites New Zealand with Australia, Madagascar with Africa, and nearly exhausts the depth of the more or less land-locked seas which lie between Australia and Asia, Africa and Europe, South and North America, and of the seas situated within the Arctic and Antarctic Circles. The Cape de Verde Islands and the Canaries belong to Africa, Madeira to Europe, and less than 500 fathoms divide Norway from Spitzbergen.

Depths from 100 to 1000 fathoms may be considered as shallow in comparison with the prevailing depths from 2000 to 3000 fathoms of the principal oceanic basins, and sufficient to establish a connection between islands and continents, the more so as we generally find one or more islands occupying the intervening space, thus betraying the common link between them.

The result of this examination is that all the larger land masses compose an area of elevation, which, after nearly completing the circuit of the world in the latitude of the Arctic circle, subdivides itself into two parts, an eastern and a western one—the former embracing Europe, Africa, Asia, and Australia, the latter North and South America. In a similar manner the different oceans combine into an area of depression, which, after making the circuit of the world along the parallel of lat. 60° South under the name of the Southern Ocean, divides itself into three large basins, respectively designated as the Pacific, the Atlantic, and the Indian Oceans. Thus the two elements, land and water, starting from opposite hemispheres, extend their arms across the Equator, holding each other in close embrace, like two champions wrestling for the mastery of the world.

DEPTH OF THE OCEAN.—A comparison of the deep-sea soundings obtained up to present date shows that, if we omit the seas situated beyond the parallels of lat. 60° N. and lat. 60° S.—no depths exceeding 2000 fathoms having as yet been ascertained beyond these latitudes—the average depth of the ocean between these parallels may be estimated at about 2500 fathoms, or more roughly at three English miles, and the average depth of all seas on the surface of the globe at probably two miles.

Contrary to the ideas formerly entertained of the enormous depth of the ocean, the soundings of H.M.S. "Challenger," S.M.S. "Gazelle," and of the U.S.S. "Tuscarora" and "Gettysburg," indicate that depths of five miles, or over 4000 fathoms, are but seldom met with, and are as exceptional as heights of the same amount on land.

The greatest depth ascertained in the Atlantic was found by H.M.S. "Challenger," in lat. 19° 41' N., long. 65° 7' W., about eighty miles north of the island of St. Thomas, in the West Indies. It is 3875 fathoms, or about four and a-half miles. In May, 1876, the "Gettysburg" found 3593 fathoms in lat. 19° 30' N., long. 65° 5' W., or only eleven nautical miles south of the

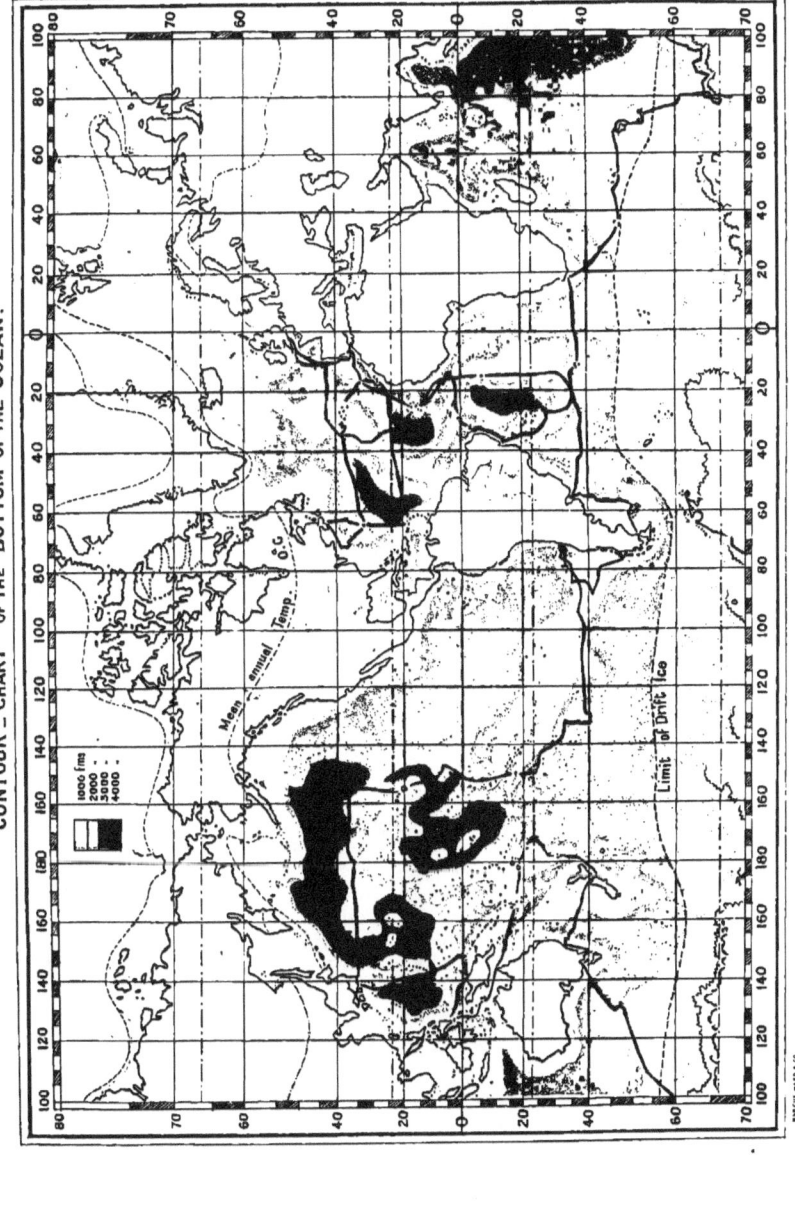

Challenger sounding, and 3697 fathoms in lat. 21° 53′ N., long. 65° 9′ W., about 132 nautical miles north of the same sounding. A depth of 3370 fathoms was obtained by the American ship in lat. 25° 47′ N., long. 65° 0′ W., which shows that the deepest area in the Atlantic is placed to the northward of the Virgin Islands, and extends over 400 miles along the meridian of 65° W.

The greatest depth observed in the Indian Ocean was discovered by the "Gazelle" in May, 1875. Two soundings of 3020 and 3010 fathoms were taken in the eastern extremity of this ocean between the north-west coast of Australia and the line of islands extending from Java to Timor.

The greatest of all depths of which we have reliable evidence was found by the "Challenger" on the 23rd March, 1875, in the comparatively narrow channel which separates the Caroline Islands from the Mariana or Ladrone Islands. This sounding is situated in lat. 11° 24′ N., long. 143° 16′ E., and amounts to 4575 fathoms, or about five miles and a quarter. Several soundings exceeding 4000 fathoms were obtained by the "Tuscarora" to the eastward of the islands of Niphon and Yezo, and another close to the most westerly of the Aleutian Islands. Two of these soundings are over 4600 fathoms, but, as it appears that no sample of the bottom was brought up, there is no evidence of the latter having been reached. H.M.S. "Challenger," shortly after her departure from Yokohama, sounded in 3950 and 3625 fathoms, and in doing so seems to have just touched the southern border of this deep but narrow area of depression, which runs parallel to the eastern coasts of Japan and the Kuril Archipelago as far as the entrance to the Behring Sea (Plate 2).

It will be observed that the above exceptional depths in the Atlantic, Indian, and Pacific Oceans are not placed, as one might be inclined to conjecture, in or near the centre of these oceanic basins, but, on the contrary, upon their confines, and in close proximity to the land. This remarkable circumstance suggests the idea that such areas of maximum depression may be the effect of a sinking of the bottom of the sea in compensation for an upward movement of the land in their immediate vicinity.

CONFIGURATION OF THE SEA-BOTTOM.—Just as the results of the recent soundings have rendered the existence of depths from six to nine miles, as formerly reported, highly improbable, so have they modified our ideas of the shape of the sea-bottom. The latter was generally represented as a repetition of the dry land with its combination of mountain, valley, and plain, a view not a little encouraged by the necessarily exaggerated scale of the oceanic sections which have appeared in print. The vertical scale is frequently from twenty-five to a hundred times in excess of the longitudinal distance, since otherwise it would be difficult to represent graphically a rise or fall of a few miles in distances measuring several hundred miles. No doubt the sea-bottom within a short distance of the shore naturally forms a continuation of the leading features of the adjoining land. Thus a large plain or a low-lying country will, as a rule, continue its almost level slopes to a considerable distance out to sea, whilst a range of hills or a chain of mountains often extends its steep inclines below the surface of the water. A comparison of the eastern and western coasts of the British Islands will afford a ready illustration of the above remarks, and examples abound in every part of the world. With regard, however, to the more central portions of the bottom of our oceanic basins, this conception of steep slopes and abrupt changes of level within short distances is not borne out by the form of the numerous oceanic sections which have been surveyed. Those who have followed day after day the results of sounding operations in mid-ocean, along sections measuring

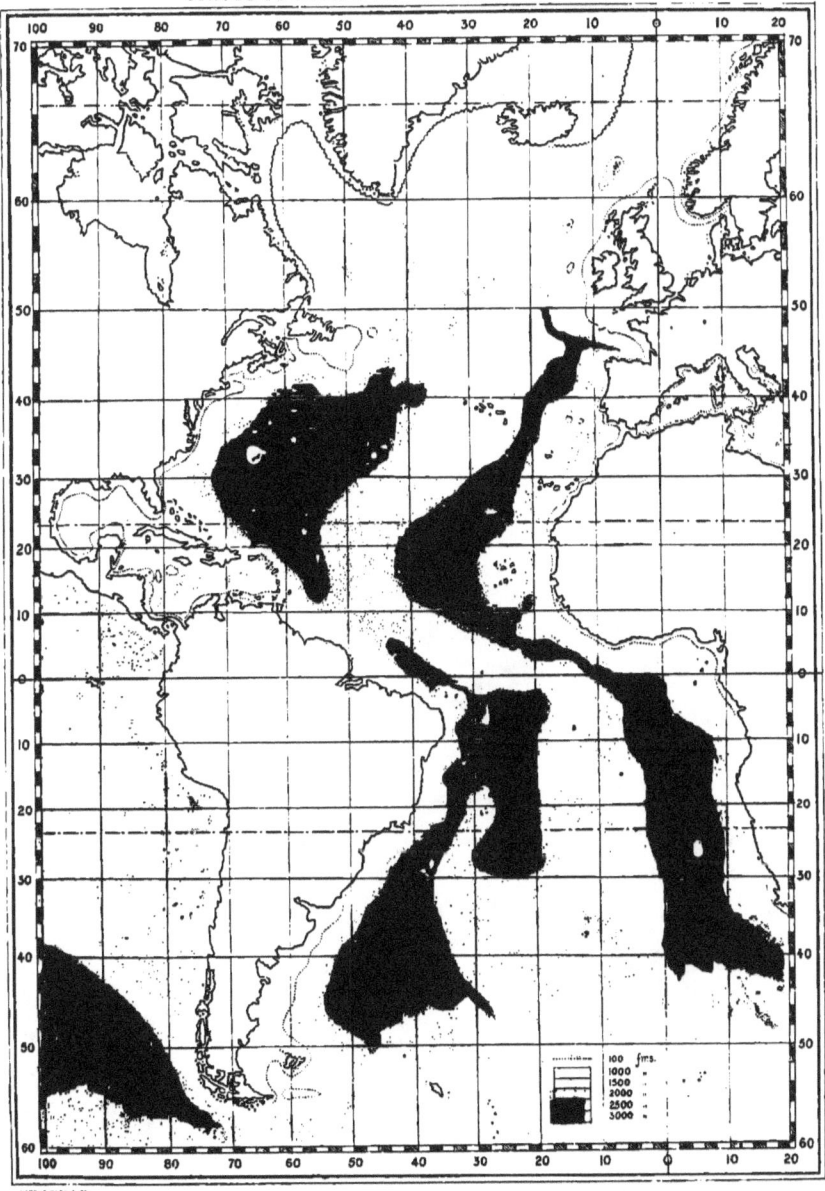

several thousand miles, will easily recall the to them familiar fact that, with rare exceptions, and those chiefly occurring in the vicinity of land, the result of one day's sounding gave a tolerably approximate idea of the depth to be encountered on the following day. The alteration of level in mid-ocean between two points as much as a hundred miles apart is generally so slight that, to an observer standing at the bottom of the sea, the latter would appear a perfect plain. Thus the bottom of our larger oceanic basins is composed of gentle undulations rising and falling from a few fathoms to two or three miles, in distances extending over many hundred miles. This view accords with the experience of the geologist, who finds that the bulk of the dry land consists of sedimentary strata originally laid down in a horizontal, or nearly horizontal, position at the bottom of the sea, and there can be little doubt but that the depths of the ocean are at the present time the scene of the formation of sedimentary strata which some day may be converted into dry land, and contain embedded in their folds traces of the animal life with which they abound.

THE BASIN OF THE ATLANTIC.—One of the most remarkable results in connection with the exploration of the sea is the discovery of several extensive submarine plateaux, which interrupt what was until lately supposed to be an unbroken waste of fathomless abyss. One of these plateaux traverses the Atlantic Ocean in its whole length from north to south, repeating in its form the S-shaped contour of the eastern and western shores of this ocean (Plate 3). After attaching itself by its northern end to the plateau which connects Europe and Iceland, and separates the Atlantic from the Arctic basin, it runs southward towards the Azores, and, gradually contracting in width, sweeps round towards St. Paul's Rocks. Reduced, comparatively speaking, to a narrow ridge, it follows the line of the Equator as far as the meridian of Ascension

Island, where, resuming its southward course, it widens out until in lat. 30° S. it occupies nearly half the space between South America and Africa, uniting the island of Ascension with St. Helena in the east, Trinidad in the west, and the group of Tristan d'Acunha and Gough Island at its southern end. In the absence of soundings to the southward of Gough Island, it is difficult to form an opinion as to whether this plateau connects itself by its southern end with the Antarctic plateau, or whether it is separated from the latter by an area of depression extending along the parallel of lat. 50° S. from the Falkland Islands to the meridian of the Cape of Good Hope. Certain indications derived from the nature of the currents, and from the deep-sea temperatures observed in that region of the South Atlantic, are in favour of the latter hypothesis.

Considerable portions of this plateau are within 1500 fathoms, or a mile and a-half, of the surface of the sea, and most of the islands are of volcanic origin. An extinct volcano, 8300 feet in height, forms the island of Tristan d'Acunha; Ascension Island rises to 2800 feet, and the summit of Pico in the Azores to 7600 feet above the level of the sea. The northern end of the plateau joins, as already stated, the plateau of Iceland with its still active focus of eruption.

By this central plateau, the Atlantic Ocean is divided into two longitudinal areas of depression or channels, one following the shores of North and South America, the other the shores of Europe and Africa. The depths vary from 2000 fathoms to nearly 4000 fathoms, the average depth being about three miles. The deepest portion of the eastern channel is situated to the westward of the Cape de Verde Islands, and forms an area of depression of over 3000 fathoms (Plate 3). In the western channel there are two such depressions, one placed between the Antilles, Bermudas, and the Azores, the other between Cape St. Roque, Ascension, and Trinidad. They are divided from each other in lat. 10° N. by a submarine elevation, which apparently connects the central plateau with the South American continent. The supposed existence of this dividing ridge is founded, not so much upon soundings, which are very few in this part of the North Atlantic, as upon the difference of bottom-temperature observed in the two depressions. The higher temperatures ascertained by the "Challenger" at the bottom of the basin north of the Equator seem to indicate that the inflow of cold water from the southern basin is arrested. The distance between the observing-stations, amounting to about 20° of lat. however, is sufficiently great to justify the conclusion that the difference of the bottom-temperature in the two areas of depression, which does not exceed 1° C. may be due to a difference of latitude, since the gradual increase from South to North of the bottom-temperature is one of the characteristic features of the Atlantic.

THE BASIN OF THE INDIAN OCEAN.—The soundings taken in this ocean prove the existence of a submerged plateau on the limit between the Indian Ocean and the Southern Ocean. It rises in many parts to within 1500 fathoms of the sea-surface, and forms the common foundation of all the islands situated in this part of the world—viz., Prince Edward Islands, the Crozet Islands, the Kerguelen group, the Heard Islands, and the islands of St. Paul and Amsterdam. The origin of all these islands is probably volcanic. The plateau occupies the space comprised between the meridians of 35° and 80° long. E., and the parallels of lat. 35° and 55° S., and the soundings obtained by H.M.S. "Challenger" between the Heard Islands and the Antarctic Circle establish a connection with the Antarctic plateau, of which the above plateau seems to be an extension towards lower latitudes.

The main basin of the Indian Ocean, with an average depth

of over 2500 fathoms, stretches from the meridian of the Cape of Good Hope towards the angle between Java and north-western Australia, where it attains its greatest depths, forming a depression of over 3000 fathoms. It communicates with the Arabian Sea by two narrow channels, situated north and south of the Chagos Archipelago, being nearly cut off from that sea by a line of islands and shallow soundings, which connect Africa, Madagascar, Bourbon and Mauritius, the Chagos Islands and the Maldivh Islands, with the Asiatic continent. The 2500-fathom line does not enter the Bay of Bengal, but it penetrates, as we have just mentioned, into the Arabian Sea, as far north as the latitude of Cape Guardafui. The 2000-fathom line forms the Bay of Bengal, and also defines the limits of a basin situated between Madagascar, Mauritius, the Seychelle Islands, and the shallow banks which unite the latter with Mauritius. The 1000-fathom line stops outside the Mozambique Channel, the Red Sea, and the Persian Gulf. The 2500-fathom area of the Indian Ocean crosses the parallel of lat. 40° S., between St. Paul and Amsterdam Islands and Cape Leeuwin in Australia, and forms, between the south coast of Australia and the forty-fifth parallel, an area of depression which extends beyond the southern end of Tasmania, includes the deepest portion of the basin between New South Wales and New Zealand, and probably communicates with the depths of the Pacific by a channel situated off the southern extremity of New Zealand.

THE BASIN OF THE PACIFIC OCEAN.—If we divide the Pacific Ocean into an eastern and a western half by a line passing from Honolulu to Tahiti, or by the meridian of long. 150° W. (Plate 2), we observe a remarkable contrast between the two portions thus formed. While the eastern half, extending towards America, presents a vast unbroken sheet of water, almost devoid of islands, the western half, towards Asia and

Australia, and enclosed between the parallels of lat. 30° N. and lat. 30° S., is composed of a labyrinth of seas, separated from each other by chains of islands, the projecting points of numerous submarine ridges. Although extensive tracts in the Pacific Ocean remain as yet untouched by the sounding-line, the observations made by the "Challenger," the "Gazelle," and the "Tuscarora," along several sections which traverse the length and breadth of this ocean, enable us to form an idea of the general contours of its bottom. From the shores of North and South America, the depths of the eastern half of the Pacific gradually increase until, upon the line between Honolulu and Tahiti, they attain 3000 fathoms. The latter depth forms extensive areas of depression in the western half of this ocean, and increases to 4000 fathoms in the already described hollow extending along the Japanese and Kuril Islands towards the entrance of the Behring Sea. Thus the idea formerly entertained of the inferior depths of the Pacific in comparison with the Atlantic, founded apparently upon the large number of islands scattered over its surface, is proved to be erroneous. Many of these islands, especially in the north-western half, rise immediately from depths of 3000 fathoms and more.

In the south-eastern portion of the Pacific there are indications of the existence of a submerged plateau connecting the Society Islands, the Low Archipelago, the Marquesas, and the intervening islands of Easter Island and Juan Fernandez with the coast of Chili and Patàgonia. H.M.S. "Challenger," after leaving Tahiti, seems to have sounded along the southern edge of this plateau down to the parallel of lat. 30° S., and as far as long. 140° W. Thence, running south towards the fortieth parallel, the ship entered the area of depression discovered by the "Gazelle," but in her eastward course along the latter parallel she once more touched the plateau in about 113° long. W., with a sounding of 1600 fathoms. In long. 94°

W., she crossed the apex of a plateau which rises to less than 1500 fathoms from the surface, and the base of which, extending from Juan Fernandez to Magellan Strait, attaches itself to the South American continent. It seems, therefore, as if an almost uninterrupted area of elevation crossed the whole basin of the Pacific in a north-westerly direction from Patagonia to Japan. The tendency of most of the submerged ridges of this ocean to follow the same direction has been frequently commented upon, and, as is the case with the submerged plateaux of the Indian and Atlantic Oceans, their association with centres of volcanic activity is equally evident.

The track of S.M.S. "Gazelle" in the South Pacific lay to the westward and southward of that of the "Challenger," and her soundings have proved the existence of an area of depression with depths of from 2600 to 2900 fathoms, bounded on the west by New Zealand, the Kermadec group, the Friendly Islands, and the Samoan Islands, and on the north by the Cook Islands, and the Tibuai or Austral Islands. It extends with lessening depths eastward towards Patagonia, along the southern edge of the above-described plateau, and probably communicates with the deep areas to the northward in the space between the Samoan group and the Society Islands. This southern area of depression, however, may be considered as belonging not so much to the Pacific as to the Southern Ocean.

A line passing from Kamtschatka over Japan, the Ladrone, Caroline, Marshall, Gilbert, Ellice, Samoa, Tonga, and Kermadec Islands to New Zealand, divides the main basin of the Pacific, of an average depth of 3000 fathoms, from the much shallower seas lying to the westward, and may possibly have formed the coast-line of a large continent which existed at a remote epoch in the history of the surface of our planet, and the boundaries of which have since been driven back to the present confines of Asia and Australia.

THE BASIN OF THE SOUTHERN OCEAN.—This ocean, which makes the circuit of the world along the parallel of lat. 60° S., in length equal to half the circumference of the Earth at the Equator, may be considered as occupying the space between the Antarctic Circle and the parallel of lat. 40° S. Owing to the limited number of soundings as yet obtained within its limits, we can only form a general idea of the distribution of its depths.

The boundary-line of the fortieth parallel, which separates the Southern Ocean from the Pacific, Atlantic, and Indian Oceans, is occupied, as has been shown in the previous pages, alternately by areas of depression, with depths ranging from 2500 to nearly 3000 fathoms, and by areas of elevation, or submarine plateaux, approaching to within 1500 fathoms of the sea surface. On the side of the Pacific, we have the deep area explored by the "Gazelle" stretching up towards the Samoan Islands, and the submerged plateau between Tahiti and Patagonia. On the side of the Atlantic, we find the southern end of the central Atlantic plateau, with the island of Tristan d'Acunha and Gough Island, flanked on the east and west by the deep areas explored by H.M.S. "Challenger." Upon the limit of the Indian Ocean, we observe what may be called the Kerguelen plateau, extending from Marion and Prince Edward Islands to St. Paul and Amsterdam Islands, as proved by the soundings both of the "Challenger" and of the "Gazelle," and the deep area, an extension of the main basin of the Indian Ocean, which passes to the southward of Australia and New Zealand, as established by the soundings of the officers of both ships. The plateaux on the border of the Atlantic and Indian Oceans may turn out to be mere extensions of the Antarctic plateau, whilst the deep areas leading to the three oceanic basins may be considered, in accordance with the observed bottom-temperatures, as the main channels by which the cold water of the Antarctic region flows northward into the Pacific, Atlantic, and Indian

Oceans. With regard to the general distribution of depth in the Southern Ocean, its bottom appears to rise gradually from nearly 3000 fathoms at the fortieth parallel (with the exception of the intervening plateaux) to little over 1500 fathoms at the Antarctic circle. There are also indications of an area of depression, of an average depth of 2000 fathoms, making the circuit of the globe between the parallels of 50° and 60° lat. S. The whole surface of the Southern Ocean is strewn with masses of floating ice, some of them forming islands many miles in extent, and rising from 100 to 300 feet above the level of the sea—an imposing spectacle, but fraught with much danger to the navigator in these regions. It is this central ocean which supplies the masses of cold water that fill up nearly two-thirds of the total depth of the Atlantic, Pacific, and Indian Oceans.

THE BASIN OF THE ANTARCTIC SEA.—We are indebted to Sir James Ross for the only soundings procured within the Antarctic circle. They are situated in the wide inlet, discovered by that illustrious navigator in the year 1840, which extends along the meridian of New Zealand, and terminates at the foot of Mount Erebus and Mount Terror. These soundings, which are all under 500 fathoms, viewed in combination with the above-mentioned gradual rise of the bottom of the Southern Ocean towards the Antarctic Circle, justify the assumption that the seas included within the latter do not exceed 1500 fathoms in depth, their average depth probably falling below this estimate. The extensive formation of ice in this region, as well as the numerous indications of land reported by the daring sailors who have penetrated so far south, suggest the hypothesis of the existence, if not of an Antarctic continent, at all events of a considerable extent of land, rising in the mountain ranges and volcanoes of Victoria Land to 10,000 and 15,000 feet above the level of the sea.

THE BASIN OF THE ARCTIC SEA.—It has been already observed in the commencement of this chapter that the region enclosed within the Arctic Circle forms an area of depression, almost completely surrounded by the land-masses of the great eastern and western continents. A shallow strait of less than fifty fathoms in depth connects it with the Pacific Ocean, and it is separated from the depths of the Atlantic by the plateau between the British Islands and Iceland, which rises to within 500 fathoms of the sea-surface. Greenland is probably the largest land-mass belonging to this basin, and next in importance we have the group of Spitzbergen, of Franz Joseph Land, discovered by the Austrian expedition; Novaya Zemblya, the Liakhov Islands, Kellett Land, off Behring Strait, discovered by the Americans in 1867; and finally the extensive archipelago, a continuation of the American continent, the extreme northern limit of which has been recently determined by the officers of the English Polar Expedition, commanded by Captain Sir George S. Nares.

The few soundings taken within the Arctic Circle leave much to conjecture, but we are tolerably safe in stating that the average depth of the Arctic basin is probably under 1000 fathoms. The immense plains of Northern Asia and America seem to continue beneath the surface of the Arctic Sea, as indicated by the numerous islands which skirt the coasts of these continents, and the greatest depths to be found inside the Arctic Circle are probably confined to the basin situated between Greenland and Norway, Iceland and Spitzbergen. This basin has been explored at different times, amongst others by Lord Dufferin in 1856, who collected a series of valuable temperature observations; by the Swedish ship the "Sophia" in 1868, to whom we are indebted for the only existing deep-sea soundings in the neighbourhood of Spitzbergen; and in the course of August, 1876, by the Norwegian Exploring Expedition on board the "Vöringen." The latter obtained a sounding of 1800 fathoms about half-way between Iceland and Norway. As a general result of the soundings taken in this basin, it appears that an area of depression of over 1000 fathoms in depth extends southward from the strait between Spitzbergen and Greenland, in lat. 80° N., as far as the Færoe Islands, and occupies the central part of the space between Greenland, Iceland, and Norway, gradually shelving up towards their shores. The volcanic island of Jan Mayen rises from the bottom of this area to a height of 6870 feet above the sea-level. The only deep-sea communication between this basin and the Atlantic seems to be effected by Denmark Strait, the unexplored space between Iceland and Greenland. According to the temperature-soundings of the "Vöringen," a mass of cold water under 0° C. fills up this basin to within 400 fathoms of the sea-surface on the Norwegian side, and to within 200 fathoms towards the east coast of Iceland. An extension of this cold stratum was discovered by H.M.S. "Lightning" and "Porcupine" in the 500-fathom channel between the Færoe Islands and the British plateau.

SURFACE – CURRENTS and SURFACE – TEMPERATURES.

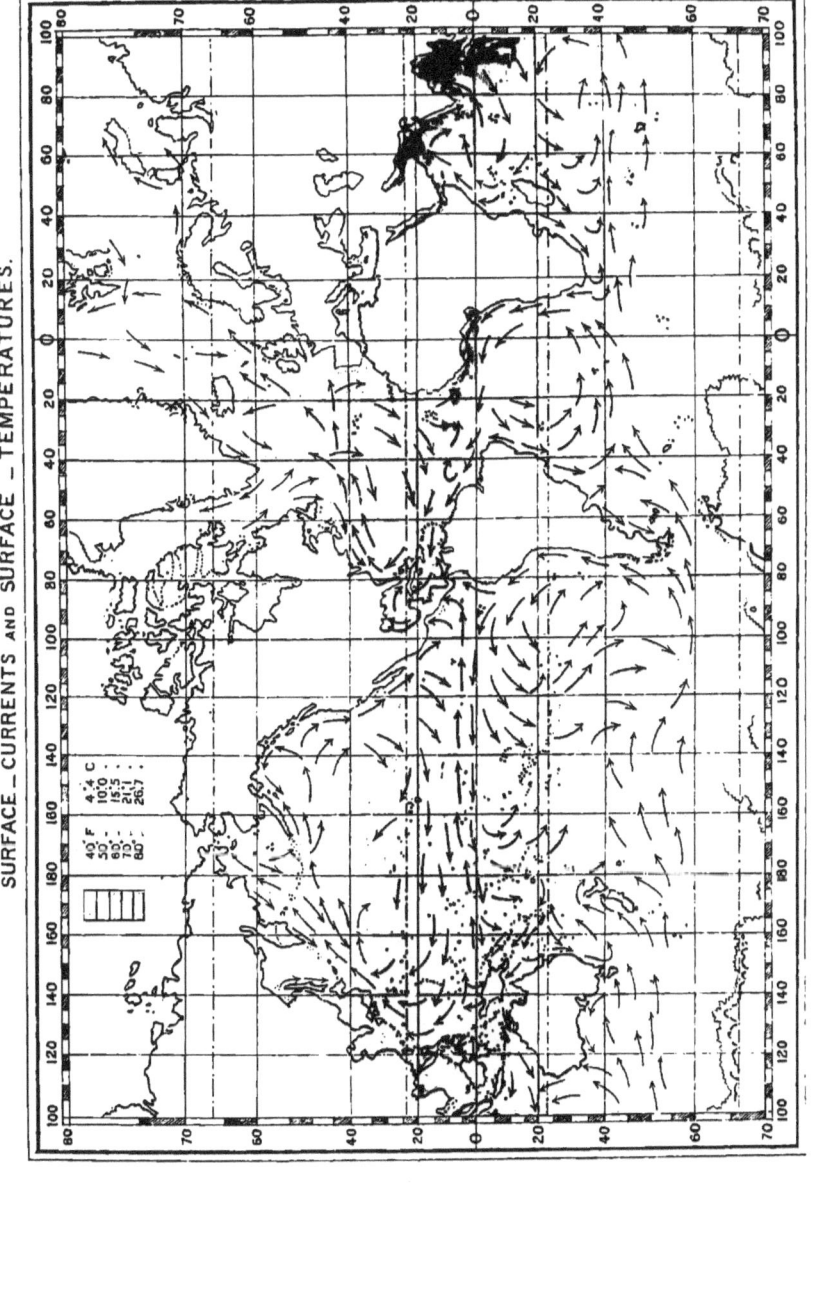

CHAPTER II.

TEMPERATURE OF THE OCEAN.

Surface Temperature—Deep-Sea Temperature—The Miller-Casella Thermometer—Serial Soundings and Temperature Curves—Deductions from the Curve.

SURFACE TEMPERATURE.—The temperature of the ocean depends mainly on three conditions—latitude, currents, and the season of the year. Owing to the unequal exposure of the different portions of the spherical surface of our planet to the rays of the sun, the amount of solar heat received gradually diminishes from a maximum between the tropics to a minimum in the polar regions. We find, in consequence, that the temperature of the surface-layer of the ocean decreases as we proceed from the Equator towards the Poles, slowly at first between the tropics, more rapidly in the temperate zones, until it falls to zero and even below zero before we reach the Arctic and Antarctic Circles (Plate 4).

In the absence of any other disturbing cause, the isotherms, or lines of equal temperature, would therefore form a system of lines running parallel with the Equator. This, however, is found not to be the case. Warm currents flowing from the tropical towards the polar regions, and cold currents issuing from the ice-bound confines of the Arctic and Antarctic, and penetrating into warmer latitudes, considerably interfere with the uniform decrease of temperature between the Equator and the Poles, although this decrease continues to be the leading feature in the distribution of oceanic temperature; but the isotherms, as a general rule, assume an oblique direction, as shown in the annexed diagram (Fig. 11).

The sun, in his apparent progress from tropic to tropic,

draws the whole system of terrestrial, including atmospheric, temperatures after him from one hemisphere to the other; but the water of the ocean, absorbing heat much more slowly and retaining it much longer than the atmosphere, is less subject than the latter to those extremes of temperature caused by the change of the seasons. A range of 5° C. or 10° F., reduced to a few degrees between the tropics and within the polar circles, measures the difference between the mean temperature of the surface-layer of the sea in the warmest and in the coldest months of the year.

The alteration of the seasons, if it does not cause a wide difference between the winter and summer temperatures of the open ocean, considerably affects the distribution of oceanic temperature not only in the surface-layer, but also in the lower strata. Both equatorial and polar currents present considerable variations in volume, velocity, and direction according to the season of the year; the warm currents acquiring a preponderance towards the end of summer, the cold currents towards the end of winter. There is thus a perpetual contest between polar and equatorial currents, especially perceptible in the latitudes where they meet, and which, on that account, may be called the critical latitudes, of which more hereafter.

The total range of oceanic temperatures embraces about 36 degrees of the Centigrade scale, commencing at $-3°.67$ C., the freezing-point of salt water, and rising above 32° C. The highest observed surface-temperatures do not belong to the Equator, but are found only in a chain of more or less land-locked seas ranged along the parallel of lat. 20° N. They are the Carribean Sea and the Gulf of Mexico in the Atlantic, the Arabian Sea, with the Red Sea and the Persian Gulf, and the Bay of Bengal in the Indian Ocean, the China Sea and the numerous smaller seas extending down to Torres Strait, and the large basin between the Philippine and Ladrone Islands in the Pacific. The surface-temperature of these seas in the warmest part of the year ranges above 28° C., and in the Red Sea rises to 32° C. The highest surface-temperature observed on board H.M.S. "Challenger" was 31°.1 C. (88° F.), and was recorded on the 21st October, 1874, in the Sea of Celebes, in lat. 4° 14' N., long. 124° 18' E. The lowest, of −2°.8 C. (27° F.), was registered on two occasions, on the 18th and 24th February of the same year, in about lat. 65° S., when surrounded by icebergs; as the temperature of the water a few hours before and after these observations was several degrees higher, the ship must have passed through what may be called pools of water of this low temperature.

One of the most remarkable examples of the influence of oceanic currents upon the distribution of temperature occurs in the North Atlantic and the adjoining portion of the Arctic basin. The cold current which flows down along the east coast of Greenland, and, after rounding Cape Farewell, joins the Labrador current, carries the western end of the isotherm of 0° C. down to the latitude of Newfoundland, while its eastern extremity is advanced as far as lat. 80° N. by a warm current which, after bathing the west coast of the British Islands and of Norway, penetrates into the North Polar regions along the coast of Spitzbergen. Thus this isotherm forms an oblique line extending from Newfoundland over Iceland and Jan Mayen to the northern end of Spitzbergen. The effect of the above-mentioned warm current, generally considered an extension of the Gulf Stream current, upon the climate of the British Islands is well known. In its absence, or if its place were occupied by a polar current—which probably was the case in a former period in the history of our planet—the climate of these islands would resemble the actual climate of Labrador.

If we consider the narrow limits of temperature, including but a few degrees of the thermometric scale, which determine the presence or absence of certain plants and animals, the power of adaptation more or less inherent in all living organisms, as well as the accumulative tendency of climatic conditions, the influence of oceanic currents upon climate, and hence indirectly upon the development of the fauna and flora in any given region, seems to afford an ample basis to account for the former presence of tropical species in latitudes now subjected to the rigours of a cold climate, and for the existence in past geological times of arctic forms in regions at present belonging to the temperate zone. In endeavouring to explain these anomalies of climate, it appears hardly necessary to go in search of vast cosmic changes, such as an alteration in the position of the terrestrial axis, a diminution in the amount of solar heat, a gradual cooling of the earth's crust, which, if they have occurred—and we have as yet no positive evidence in their favour—would imperil the existence of all life on our planet; while we have, close at hand, an agency whose effect upon climatic conditions may be said to be a matter of daily experience, and which is sufficiently powerful to establish, in almost any region on the earth's surface, the small difference of temperature which is a decree of life or of death to numerous animal and vegetable organisms.

DEEP-SEA TEMPERATURE.—We are indebted to the officers of the United States Coast Survey for the first systematic investigation of the conditions of temperature below the sea-surface. Their operations, extending over fourteen years, from 1845 to 1849, determined the course of the two great currents which, under the name of the Gulf Stream and the Labrador Current, were known to flow along the United States coast between Nantucket Island and the Gulf of Mexico. Although not provided with the more perfect appliances and instruments placed at the disposal of recent explorers, their observations furnished ample evidence of the existence below the sea-surface of vast bodies of moving water thousands of miles in length, hundreds of miles in breadth, and hundreds of fathoms in depth, now flowing side by side, now passing one above the other. It was shown, for the first time, that a polar current may prolong its course along the bottom of the sea into tropical latitudes, as it was known before that warm currents of equatorial origin traverse the temperate zone and extend their life-giving influence into the polar regions.

Apart from the obvious difficulties that stand in the way of carrying on a series of scientific observations on the high seas with delicate instruments, the chief desideratum in these earlier experiments was felt to be in the possession of a thermometer constructed to register with sufficient accuracy the temperature of the water at great depths, and to minimise the effects of the enormous pressure, amounting to about a ton on the square inch for every thousand fathoms. This pressure, exercised on the bulb of an ordinary thermometer, if it did not effect the total destruction of the instrument, always caused the latter to record a higher temperature than that of the water at the depth to which it had been lowered, the difference increasing with the depth to as much as $5°$ C. for 2000 fathoms, and often more.

The want of an efficient deep-sea thermometer, which more or less vitiates all the observations of earlier explorers, was especially experienced during the cruise of H.M.S. "Lightning" in 1868, the first expedition fitted out for sounding and dredging purposes by the joint co-operation of the Royal Society and the Admiralty of England. Of the thermometers used on this occasion, several returned to the surface broken by the pressure to which they had been exposed, and the indications given by the rest varied so much as to render the discovery of a thermometer free from this error a matter of paramount importance for the success of future deep-sea exploration. By a happy coincidence, the desired improvement was effected at the moment when a second expedition was being prepared. In April, 1869, previous to the departure of H.M.S. "Porcupine" on her first cruise, at a meeting of the Deep-sea Committee of the Royal Society of London, Dr. W. A. Miller, V.P.R.S., suggested a simple expedient for protecting thermometers from the effects of pressure, which, ably carried out by Mr. L. P. Casella, F.R.A.S., the eminent scientific instrument maker to the Admiralty, resulted in the construction of an almost perfect instrument for recording the temperature at great depths. Since all the observations made on board H.M.S. "Challenger," during her cruise round the world, were obtained with the help of this instrument, now known as the Miller-Casella thermometer, a description of it in these pages may not be out of place.

THE MILLER-CASELLA THERMOMETER.—It will be seen from the accompanying figure that this thermometer is designed to register the maximum and minimum temperature of the water to which it is exposed during its descent from the surface of the sea to the bottom. For this purpose the glass tube is bent in the shape of U, each arm of the tube terminating in a bulb. The larger bulb, A, is surrounded by another bulb, B, and about three-fourths of the space between the two bulbs is filled with alcohol. It is by the addition of this outer bulb, B, that the protection of the instrument from the effects of pressure is secured. On immersion, the outer bulb receives the pressure of the water, and forces the enclosed alcohol into the portion of the intervening space previously not occupied by this liquid, thus relieving the inner bulb, A. The latter is completely filled with a mixture of creosote, alcohol, and water, which rests upon the mercury contained in the bend of the tube, and also fills up the other arm and part of the bulb C. The upper part of bulb C is occupied by air, introduced, with the help of a freezing

THE MILLER-CASELLA THERMOMETER.

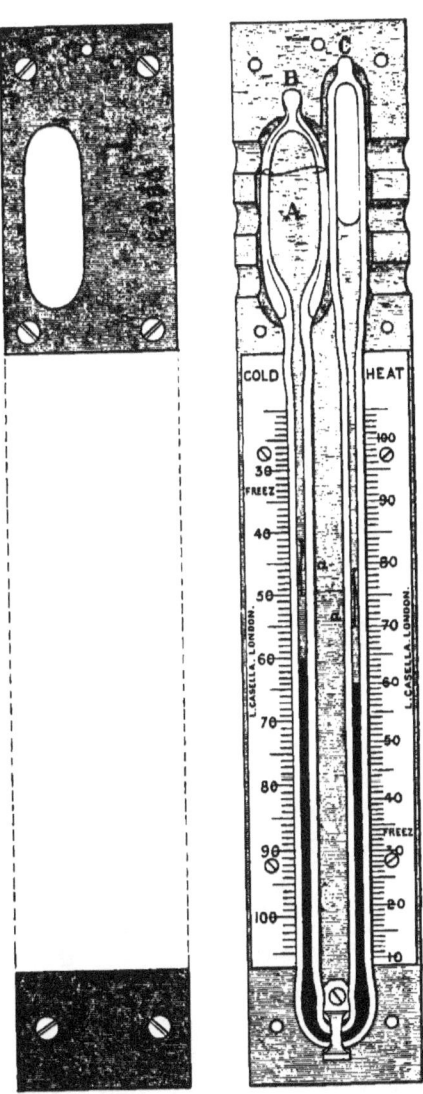

Scale 2/3 Nat. Size.

mixture, at a very low temperature in order to increase its density. It acts as a sort of elastic cushion intended to overcome the friction of the liquids in the tubes, and to assist the mercury in following the mixture when contracting under the influence of cold. The indications of the thermometer depend upon the expansion by heat and contraction by cold of the mixture contained in bulb A. When expanding, the mercury is forced down in the arm attached to this bulb and rises in the other arm towards bulb C; when contracting, the mercury falls on the side of bulb C, and rises towards bulb A. Two metal indices, *a a*, mark the maximum height which the mercury has reached in either arm, and a hair attached to each index produces the friction necessary to retain them at the level to which they have been raised. Before the thermometer is lowered into the sea, the indices are brought down upon the mercury by passing a magnet along the tube. One of the essential qualities of an instrument intended for use on long voyages is that it should be portable—a quality especially realised by Mr. Casella in the construction of his thermometer. By contracting the bore of the glass tube as much as possible, the quantity of the liquids, particularly of the mercury, has been reduced to a minimum; and the liability to accident, almost inseparable from instruments containing large quantities of this heavy substance, has thus been greatly reduced.

The cruise of H.M.S. "Challenger" afforded ample opportunities for testing the capabilities of the Miller-Casella thermometer. It resisted all the pressures to which it was exposed down to a depth of about 4000 fathoms, when some of the instruments were found to give way under a pressure of four tons to the square inch; but as depths of from four to five miles are exceptional, such accidents will be of rare occurrence. Under a pressure of three tons, equivalent to a depth of three miles, the error amounts to less than $1°$ C., whilst that of the unprotected thermometers previously in use sometimes exceeded 10° C. for the same depth; and a comparison of the data furnished by the Miller-Casella thermometer with the corresponding temperature curves shows that the mean error of all the deep-sea observations made on board H.M.S. "Challenger," not much under 10,000, is probably less than 0°.5 C. The thermometers, before being sent out, are subjected to pressures varying from one to four tons to the square inch in a hydraulic press especially designed for this purpose by Mr. Casella, and the amount of error ascertained for each instrument. When in actual use, they are enclosed in a copper cylinder perforated at both ends to allow free ingress and egress to the water. Several thermometers may be attached to the same sounding-line whilst it is being paid out, at distances of five, ten, twenty-five, fifty, or one hundred fathoms as required; and it is found that an immersion of from five to ten minutes is sufficient to secure the desired record of temperature.

SERIAL SOUNDINGS AND TEMPERATURE CURVES.—In order to ascertain the distribution of temperature at any selected station upon the ocean, it is necessary to obtain as many observations at various depths between the surface and the bottom as circumstances will permit. The amount of time absorbed in hauling in the sounding-line from great depths, even with the help of a steam-engine, as well as the uncertain state of the weather, which may interrupt the operations at any moment, necessarily limit the number of observations which can be taken in the course of a stoppage on the high seas. However, the first experiments made of this kind establish the fact that, beyond a depth of from 1000 to 1500 fathoms, the temperature of the water decreases very slowly, often not more than 0°.1 C. in 100 fathoms; and that, beyond these depths, observations taken at intervals of 100, 200, or 250 fathoms are sufficient for all purposes. A series of observations at every 10 fathoms from the surface to 100 fathoms, every 25 fathoms down to 300 fathoms, and every 100 fathoms down to 1000 or 1500 fathoms, furnishes, with the addition of the bottom-temperature, sufficient materials for scientific enquiry; and, although it may be secured in the course of a few hours, it can only be accomplished under favourable conditions of weather. The number of serial soundings obtained by H.M.S. "Challenger" during her three and a-half years' cruise amounts to about 260, of which 120 belong to the Atlantic, 110 to the Pacific, the remainder being about equally divided between the Southern Ocean and the seas of the Indian Archipelago. An important series of soundings was taken almost simultaneously by the officers of the Imperial German frigate the "Gazelle," whose operations in the Indian Ocean and in the south-western portion of the Pacific, not visited by H.M.S. "Challenger," form a valuable supplement to the results obtained by the officers of the English expedition.

When recording the observations made at a particular station, it becomes at once apparent that a mere tabulated statement of the temperature registered at each depth is but an imperfect mode of exhibiting the results of the soundings, and recourse is had to the method of curves. It is not necessary here to insist on the evident advantages of this method of representing graphically the various stages of a phenomenon under observation, and on the facilities which it offers not only for seizing at a glance its leading features, but also for detecting the errors of the instruments employed and those of the observer himself.

To convert a tabulated statement of a series of temperature observations belonging to one station into a curve, it is sufficient to lay down a scale of fathoms, say along a horizontal line, and, at right angles to the latter, a scale of degrees of temperature (Figs. 1 to 10). At the points on the horizontal scale corresponding with the depths at which the observations were taken, perpendicular lines are drawn, and their lengths made equal to the number of degrees registered by the thermometer. Thus these lines will be longer or shorter in proportion as the temperature increases or decreases. On joining the upper end of the lines we obtain a curve of a more or less regular shape, generally assuming the form of the letter S, which is called the temperature curve of the station, and it exactly represents the rate of increase or decrease of the temperature of the water in each stratum from the surface to the bottom.

To exhibit the conditions of temperature ascertained at a number of stations belonging to a certain section of an oceanic basin, another class of curves may be constructed (Plates 6 to 20). In this case the horizontal base-line forms a scale of miles, or of degrees of latitude or longitude, giving the distance between one station and another. The vertical scale indicates the depth in fathoms, and the curves thus obtained form isotherms exhibiting by their rise and fall the decrease or increase of temperature in the different strata of the oceanic section under consideration. They also show the rate of increase or decrease, spreading out when the temperature decreases more slowly at one station than at another, and closing up when the decrease is more rapid. Instead of a scale of fathoms, the vertical scale may be divided into degrees of temperature, and the curves resulting from this arrangement will form iso-bathymetric lines, exhibiting the variation of temperature for the same depth at different stations. In the plates and tables prepared for this essay, the former arrangement has been given the preference, since, by using a scale of depths, the diagram represents an actual section of an oceanic basin, although on a scale necessarily much exaggerated.

DEDUCTIONS FROM THE CURVE.—An examination of the shape of the curve, and of the modifications which it undergoes in the different parts of the ocean, leads to certain conclusions respecting the conditions which determine the distribution of oceanic temperature, and the direction in which we must seek for a solution of the problem of oceanic circulation.

In looking over the temperature curves and sections which accompany these pages, we find *that, as a general rule, the temperature of the ocean decreases from the surface to the bottom.* One of the first surprises in store for the early observers of deep-sea temperature was the discovery that the temperature of the deeper strata, and at the bottom of the ocean, is only a few degrees removed from freezing-point. Subsequent observations have shown that, with certain exceptions, this is the case in every part of the ocean, in tropical latitudes as well as in the temperate and frigid zones, and that over some areas the temperature of the lowest strata falls even below zero. The first well-ascertained example of the presence of cold water in low latitudes is due to the soundings of H.M.SS. "Lightning" and "Porcupine" in 1868–69, in the channel between the Færoe Islands and Scotland. Over an area situated between lat. 60° and 61° N., long. 4° and 9° W., the temperature of 0° C. is reached at a depth of only 300 fathoms, descending to $-1°.4$ C. at a depth of 640 fathoms. This mass of cold water—about 120 miles long, 60 miles broad, and a quarter-mile in depth—must have come from the Arctic region, and the soundings of the Norwegian ship "Vöringen," in the course of last year, show that it is a southern extension of the mass of cold water which fills up the basin between Iceland and Norway to a depth of 1800 fathoms. A still more remarkable instance of the existence of a cold area with a temperature below zero was discovered off the mouth of the Rio de la Plata by H.M.S. "Challenger" on her return voyage in 1876. Between lat. 36° and 42° S., long. 33° and 55° W., the bottom-temperature at seven different stations was found below 0° C., varying from $-0°.3$ to $-0°.6$ C. The maximum depth found in this area was 2900 fathoms, and the zero-point was reached in 1900 fathoms (Station 323).

This cold area may be traced southwards into the Antarctic region, and northwards as far as the Equator; for at Station 112, off Cape St. Roque, the temperature at a depth of 2200 fathoms is only 0°.5 C., and at Station 110, under the Equator, 0°.9 C. at a depth of 2275 fathoms. At the two latter stations, the temperature of 5° C. is reached at a depth of little over 300 fathoms.

As early as 1859, the officers of the U.S. Coast Survey found a temperature of 4°.4 C. (40° F.) at a depth of 300 fathoms in the Strait of Florida; the water of the Gulf Stream at the surface having, at the same date, a temperature of 26°.7 C., or 80° F.

The two principal exceptions to the above-stated rule of the gradual decrease of temperature from the surface towards the bottom are to be found, one in a number of small basins which are cut off by submarine ridges or elevations from communication with the lower strata of the ocean outside, the other in the Arctic and Antarctic regions, where the rule seems to be completely inverted, the temperature of the water increasing from the surface towards the bottom.

The operations of H.M.S. "Porcupine" in the summer of 1870 in the Mediterranean, and the cruise of H.M.S. "Challenger," furnish examples of the first-mentioned exception. It is well known that the Mediterranean Sea is cut off from the depths of the North Atlantic Ocean by a submarine elevation extending from Cape Spartel in Africa to Cape Trafalgar in Spain, and rising to within 100 fathoms from the sea-surface. From a series of temperature-soundings obtained by H.M.S. "Porcupine" at seven different stations between the meridian of Malaga and Carthagena, it appears that the temperature of that part of the Mediterranean falls from a mean of 22°.6 C. at the surface to 13° C. at a depth of 100 fathoms, whence it remains stationary at the latter temperature down to the bottom, at depths varying from 162 to 845 fathoms. From this it has been inferred, *that in a basin separated from the adjoining ocean by a submarine elevation, the temperature of the water decreases from the surface down to the level of that elevation, and remains stationary from that level down to the bottom.*

The subsequent proceedings of H.M.S. "Challenger," in the Western Pacific and in the seas of the Indian Archipelago (Plates 15 and 16), leave little room for doubting the correctness of this conclusion. In the basin extending from the New Hebrides to Torres Strait (which might appropriately be called the Melanesian Sea), in the Banda Sea, Celebes, Sulu, and China Sea, the temperature of the water was found to decrease from the surface down to a certain level, and remain stationary between that level and the bottom. The Sulu Sea furnishes the most striking illustration of this hitherto unsuspected phenomenon. At the time of the two visits of H.M.S. "Challenger," in October, 1874, and January, 1875, the temperature was observed to fall from 28° and 27° C. at the surface to 10°.2 C. at a depth of 400 fathoms, and to remain stationary at the latter temperature from 400 fathoms down to the bottom, at depths of 2550 and 2225 fathoms, forming a stratum of more than two miles in depth of the comparatively high temperature of 10°.2 C., or 50°.4 F. Although we possess no complete surveys of the bottom of the seas above mentioned, a glance at a chart will show that they all are more or less land-locked basins, and, as the temperature-conditions seem to prove, probably cut off from the colder strata of the Pacific and Indian Oceans by submarine ridges rising to the level at which the decrease of temperature is arrested.

The second exception, observed both in the South Polar and North Polar regions, may require further research before it can rank as an established scientific fact. Besides the numerous difficulties which beset thermometric observations under the severe conditions of a polar climate, the limits of oceanic temperature in these latitudes are so narrow as to render additional caution necessary. There is, however, considerable agreement between the observations made by the several explorers who have penetrated into these inhospitable regions, who all assert the discovery of warmer water below the cold surface-stratum, so that the fact seems hardly doubtful; and the recent experience of the officers of H.M.S. "Challenger" in the vicinity of the Antarctic Circle points in the same direction (Fig. 8, Curve B, and Plates 12 and 13). On theoretical grounds it may be said that the existence of open water in the polar regions, in contact with an atmosphere the temperature of which is generally below freezing-point, shows that warm currents from lower latitudes must find their way into these regions. The constant melting of the ice floating in these warm currents must tend to produce layers or pools of water of a temperature near freezing-point and of a lower specific gravity, which, for a time, must remain at or near the surface, and form strata of a temperature lower than that of the strata beneath.

It will appear from the previous remarks that the distribution of temperature in the ocean depends, as a general rule, upon a constant supply of heat at the surface and a constant supply of cold at the bottom; and the temperature-curve will represent a series of gradually decreasing temperatures from the surface towards the bottom, as in Fig. 1. In those regions where the supply of heat is reduced to a minimum, as we find is the case in the higher latitudes, the stratum of cold water will be reached within a short distance from the surface (Fig. 2, Curve C, and Fig. 7), and in some parts of the ocean it may be said to occupy the whole depth of the sea (Fig. 8, Curve A). On the other hand, where, as in the cases mentioned above, the supply of cold is reduced or entirely cut off by submarine barriers, the tem-

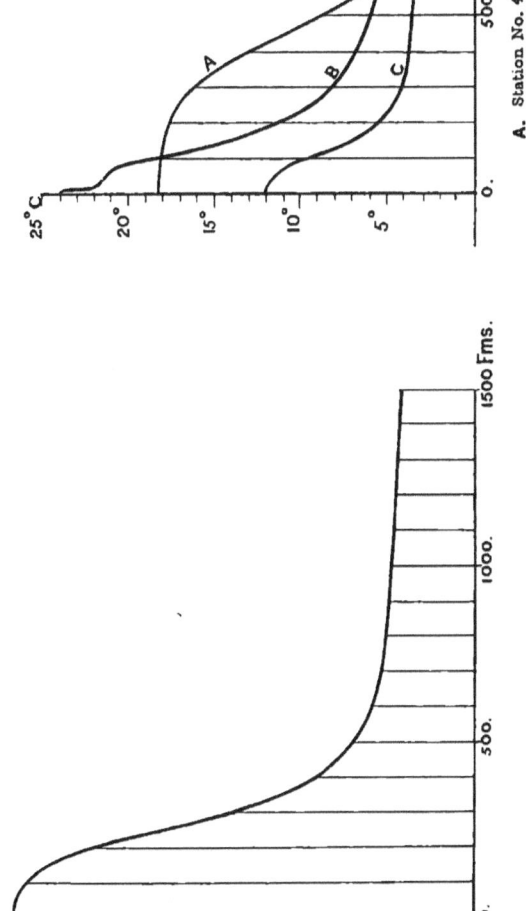

perature will remain stationary at or near the level of the obstruction, and the stratum of warm water will extend to the bottom, and thus fill up the whole space between the surface and the bottom, whatever may be the depth of the basin (Plate 16).

This constant supply of heat and of cold is effected, as is well known, through the agency of currents. The latter are by no means an exceptional phenomenon confined to certain parts of the ocean. Varying in volume and velocity until they attain the proportions of gigantic rivers flowing at a rate of several miles an hour, they occupy every part of the ocean, no part of which can be said to be in a condition of absolute rest. Combined together they form, like the currents in the atmosphere, and in intimate association with the latter, a complete system of circulation, by which the physical and chemical equilibrium of the ocean is maintained. From the principal storehouse of heat in the tropics, warm currents proceed towards the temperate and frigid zones, and return thence in the character of cold currents towards the regions of the Equator. That this is so is proved by the results of all observations made up to the present day, and it is in perfect agreement with the well-known agency of water as a storer-up and carrier of heat.

Two strata of different temperatures cannot remain in contact for any time without the formation of an intermediate stratum. This is presumably the reason why a series of deep-sea temperature observations generally assumes the shape of a curve, nowhere presenting a break or an abrupt transition from one temperature to another. The depth of this intermediate stratum will depend upon the duration of the contact. When two masses of water, one warm the other cold, move in different or opposite directions, the intermediate stratum will present a rapid transition from the temperature of one stratum to the temperature of the other, and the part of the curve which represents the intermediate stratum will form a steep incline. On the contrary, when two masses of water flow in the same or nearly the same direction, the intermediate stratum will appear in the curve as a gradual incline representing the slow increase or decrease of temperature from one stratum to the other. Hence the gradient of any part of the curve is not only the measure of the rate of increase or decrease of temperature, but also an indication of the relative motion or relative rest of the strata in contact. A low gradient expresses the presence of strata of equal or nearly equal temperature moving in the same direction, *i.e.*, at relative rest towards each other; a steep gradient indicates the existence of strata of different temperatures, and moving in different directions.

Thus, in Fig. 1, we have a warm surface-stratum of considerable thickness, the decrease of temperature in the first 200 fathoms amounting to only $2°.4$ C. The steep gradient between 200 and 500 fathoms shows that this surface-stratum moves in a direction different from that of the bottom-stratum, which, at this station, is found to rise up to within 600 fathoms from the surface. The temperature at the bottom, at a depth of 2550 fathoms, is $0°.7$ C.; at 1500 fathoms, $2°$ C.; at 600 fathoms, $2°.9$ C., or a difference of only $0°.9$ C. in 900 fathoms.

In Fig. 2, Curve A, belonging to Station 41, near the eastern limit of the Gulf Stream, the hump extending down to 300 fathoms represents a stratum of nearly uniform temperature. The latter is, at the surface, $18°.3$ C.; at 125 fathoms, $18°.0$ C.; at 300 fathoms, $17°.0$ C. In Curve B, at Station 43, in the Gulf Stream itself, all this warm water below 100 fathoms has disappeared; but a surface-stratum has been added, of a temperature rising to $24°$ C. Still further on, at Station 44 (Curve C), a short distance beyond the western limits of the Gulf Stream, we find ourselves in the midst of an Arctic current rising up to within 300 fathoms from the surface, with a

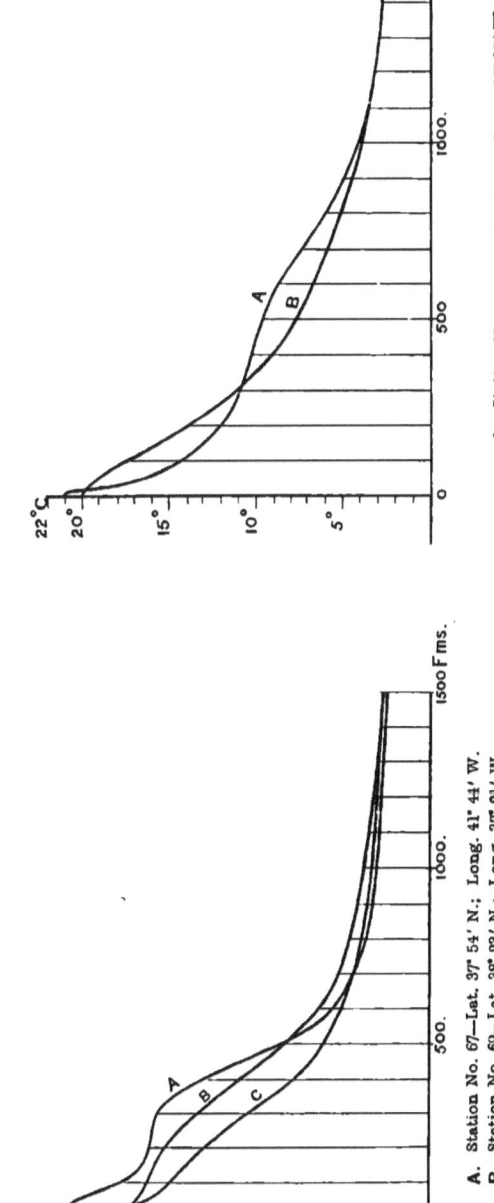

Fig. 3.

TEMPERATURES IN THE NORTH ATLANTIC.

A. Station No. 67—Lat. 37° 54′ N.; Long. 41° 44′ W.
B. Station No. 69—Lat. 35° 23′ N.; Long. 37° 21′ W.
C. Station No. 71—Lat. 35° 18′ N.; Long. 34° 43′ W.

Fig. 4.

TEMPERATURES IN THE NORTH ATLANTIC.

A. Station No. 82—Lat. 33° 46′ N.; Long. 19° 17′ W.
B. Station No. 5—Lat. 24° 20′ N.; Long. 24° 28′ W.

Fig. 5.

TEMPERATURES NEAR THE CAPE OF GOOD HOPE, AGULHAS CURRENT.

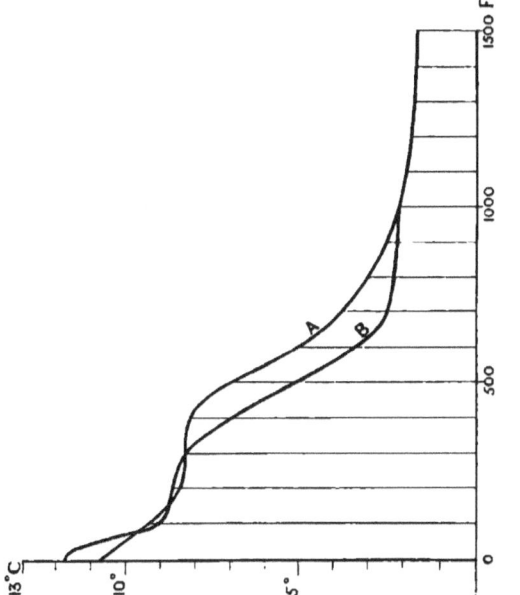

A. Station No 143—Lat. 36° 48' S.; Long. 1° 24' E.
B. Station No. 139—Lat. 35° 35' S.; Long. 16° 8' E.

Fig. 6.

TEMPERATURES IN THE SOUTH-AUSTRALIAN CURRENT.

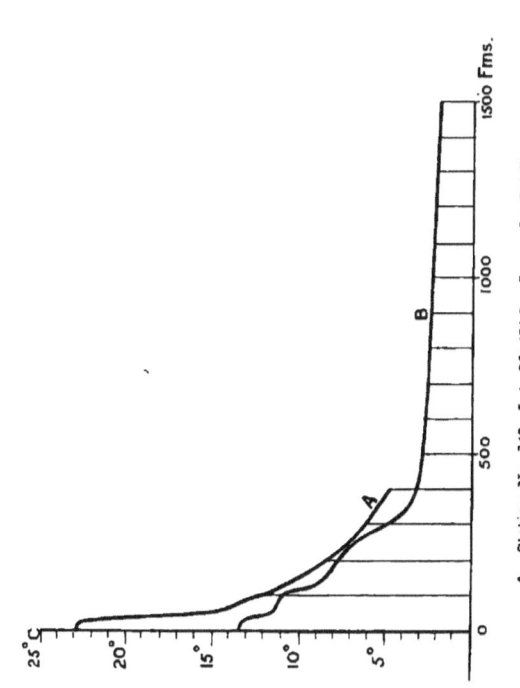

A. Station No. 159—Lat. 47° 25' S.; Long. 130° 32' E.
B. Station No. 160—Lat. 42° 42' S.; Long. 134° 10 E.

temperature of 4°.1 C. at 300 fathoms, of 3° C. at 900 fathoms, and of 2° C. at 1500 fathoms. The temperature of the surface has fallen from 24° C. to 11°.1 C. The space which contains the three stations covers 1° 20′ of latitude, and 1° 57′ of longitude—truly an extraordinary change in the distribution of temperature in so small a portion of the Atlantic.

In Fig. 3, the Curves A, B, C illustrate the gradual disappearance towards the Azores of the same stratum, 300 fathoms in thickness, which appears in Curve A, Fig. 2.

In Curve A, Fig. 4, the hump between 200 and 700 fathoms (12° C. to 7°.5 C.) marks the presence of a large current flowing between the Azores and Madeira; and the low gradients of Curve B indicate the existence, between the surface and 1500 fathoms, of various strata of gradually decreasing temperatures moving more or less in the same direction.

Curve A, Fig. 5, shows the rise of temperature caused by the Agulhas current, to the southward of the Cape of Good Hope; while Curve B of Station 139, situated but a short distance westward of the Cape, affords little or no indication of the proximity of this large current of warm water. The unusually irregular shape of Curve B betrays the existence in the vicinity of the Cape of numerous currents moving in different directions one above the other.

The two curves in Fig. 6 illustrate the temperature-conditions in the great South Australian current. H.M.S. "Challenger," after a three months' cruise, her stores of coal having run short, was precluded from establishing numerous stations on her way from the Antarctic Circle to Australia. Thus it was found, on arriving at Station 159, that the ship had already crossed the southern limit of this great current, of which, at Station 158, in lat. 50° S., long. 123° E., there had been little or no trace (Plate 12). The isotherm of 5° C., which at the latter station was reached at a depth of 200 fathoms, fell to 600 fathoms at Station 159, and rose again to 500 fathoms at Station 160, so that between Stations 159 and 160, the "Challenger" must have crossed the axis of a current about 500 fathoms deep, and from 500 to 600 miles broad. This current, coming from the Indian Ocean, flows in a south-easterly direction to the southward of Australia, and penetrates into the Antarctic region along the meridian of New Zealand.

Fig. 7 presents a section, at Station 318, of the great Antarctic current which flows as an under-current along the east coast of South America, crosses the Equator, and penetrates into the North Atlantic. At the above station it rises to within 100 fathoms of the surface. The steep gradient between the surface and 100 fathoms is due to a branch of the Brazilian current, which flows in a southerly direction towards the Falkland Islands.

Curve A, Fig. 8, furnishes a similar example of the presence of a cold stratum at the depth of little more than 100 fathoms from the surface. It is the temperature-curve of Station 147, near the Crozet Islands. Curve B illustrates the case of a cold surface-stratum, probably formed by melting ice, observed in the vicinity of the Antarctic Circle. The temperature falls from $-1°.2$ C. at the surface to $-1°.7$ C. at 50 fathoms, but rises to $-0°.8$ C. at 200 fathoms, $0°.0$ C. at 300 fathoms, and $0°.4$ C. at a depth of 500 fathoms (Plates 12 and 13).

Figs. 9 and 10 represent the conditions of temperature near the Equator in the Atlantic and Pacific Oceans. The curve of Fig. 9 belongs to Station 110, near St. Paul Rocks; the curve of Fig. 10 to Station 221, in the basin between Papua and the Caroline Islands. In the former a surface-current, retaining a nearly uniform temperature of $25°$ C. down to a depth of 30 fathoms, is joined by a steep gradient to an intermediate current which extends from 100 fathoms to 400 fathoms, the cold bottom-stratum being reached at a depth of

Fig. 7.

TEMPERATURES IN THE SOUTH ATLANTIC, ANTARCTIC CURRENT.

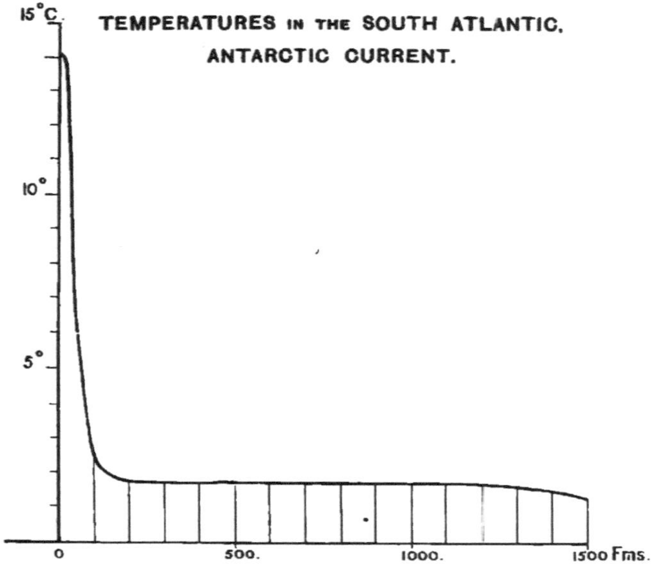

Station No. 318—Lat. 42° 32′ S.; Long. 56° 27′ W.

Fig. 8.

TEMPERATURES IN THE SOUTHERN OCEAN, ANTARCTIC ICE-BARRIER.

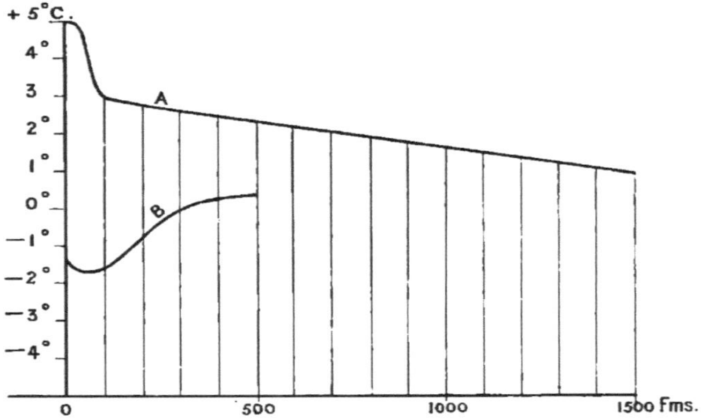

A. Station No. 147—Lat. 46° 16′ S.; Long. 48° 27′ E.
B. Station No. 153—Lat. 65° 42′ S.; Long. 79° 49′ E.

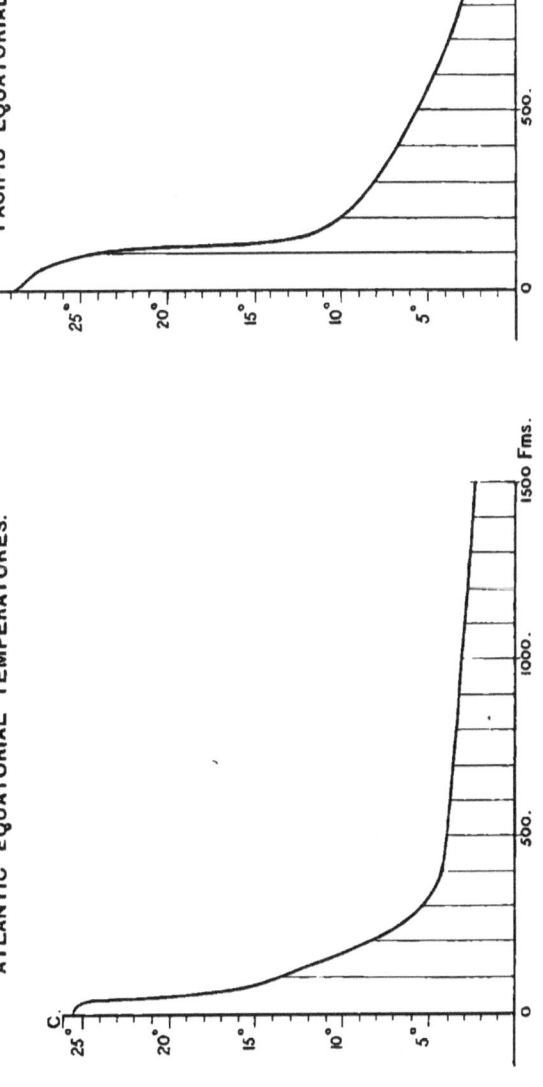

Fig. 9. ATLANTIC EQUATORIAL TEMPERATURES. Station No. 110—Lat. 0° 9′ N.; Long. 30° 18′ W.

Fig. 10. PACIFIC EQUATORIAL TEMPERATURES. Station No. 221—Lat. 0° 40′ N.; Long. 148° 41′ E.

500 fathoms with a temperature of 4° C. At the station in the Pacific, a surface-stratum, the temperature of which falls from 28°.8 C. to 28° C. at 30 fathoms, and to 26° C. at 80 fathoms, is united by a steep gradient to an intermediate stratum, which extends from 150 fathoms (11°.3 C.) to 800 fathoms (3° C.), the bottom-stratum commencing at 900 fathoms with a temperature of 2°.5 C. How soon the cold bottom-stratum is reached in the equatorial belt is one of the unexpected discoveries due to recent deep-sea exploration. In the warm seas which bathe the British Islands, a temperature of 4° C. is not registered until we arrive at a depth of 900 fathoms, and at 1500 fathoms the temperature is still 2°.5 C.

CHAPTER III.

CURRENTS OF THE OCEAN.

The Aqueous and the Aerial Oceans—Thermal Circulation—Vertical and Horizontal Extension of the Two Terrestrial Envelopes—Parallelism between Oceanic and Atmospheric Currents—Surface and Under-Currents.

THE AQUEOUS AND THE AERIAL OCEANS.—The aqueous envelope, which, as we have seen, covers about three-fourths of the surface of the solid crust of the earth to an average depth of from two to three miles, is itself surrounded by and everywhere in contact with another envelope termed the atmosphere, which forms an "aerial ocean" covering the whole surface of our planet to a depth supposed not to exceed eighty miles. Whether this aerial ocean has a well-defined surface like the aqueous ocean is a point which remains to be settled by future research. What we know is, that the density of the various strata into which it may be divided decreases so rapidly that at a height or depth of 18,000 feet, or 3000 fathoms, we have already left behind one-half of the mass of air of which it is composed. It has also been ascertained by recent observations, that the proportion of aqueous vapour—upon the presence of which in the air the agency of the atmosphere as a storer-up of heat and moisture mainly depends—diminishes with equal rapidity, and is, as far as observation goes, reduced to zero at a distance of only a few miles from the earth's surface. The thickness of the atmospheric layer, considered as a meteorological agent, may therefore be safely reduced to five miles, or even less, for the greater number of the atmospheric phenomena with which we are immediately concerned take place within a distance of from two to three miles from the earth's surface.

Ever since the movements of the atmospheric air and of the

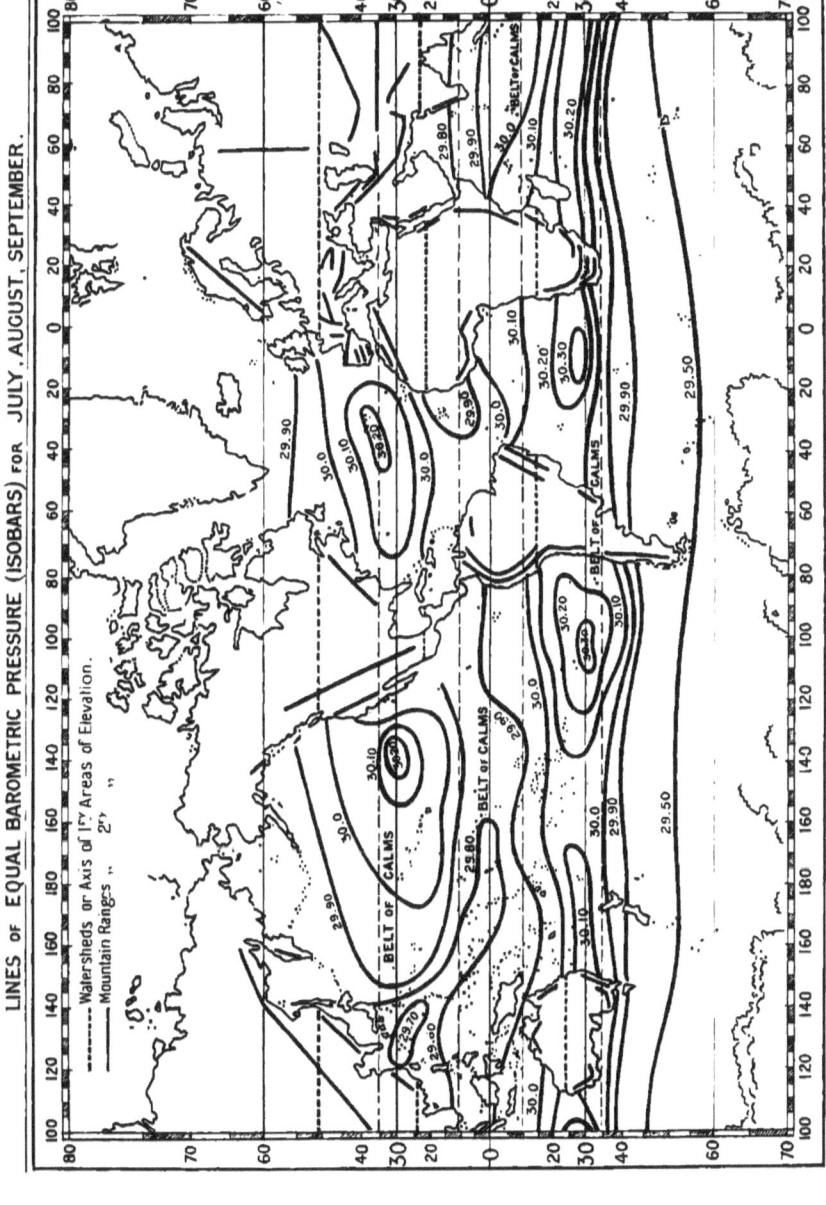

waters of the ocean have attracted the attention of the scientific observer, the resemblance between the phenomena which occur in the gaseous envelope and those observable in the aqueous envelope of our planet has frequently been pointed out. This resemblance is the obvious result of a similarity of conditions and an identity of natural laws which govern the internal economy of the two envelopes. Both are composed of fluids subject to the laws of gravity, and to the laws which direct the movements of fluids in general, their expansion under the influence of heat, their contraction under the action of cold. The equilibrium of both is constantly disturbed in consequence of the unequal distribution of solar heat between the Poles and the Equator, and is as constantly restored through the agency of currents, cold air and cold water unceasingly flowing from high towards low latitudes, warm air and warm water without intermission passing from the torrid zone into the temperate and polar regions. As recent observations have shown, in both, in the aqueous as well as in the aerial ocean, the temperature rapidly decreases from the surface towards the deeper strata (considering the stratum of the atmosphere which is in close contact with the surface of the earth as the virtual surface of the aerial ocean); and the surface-stratum of both forms a stratum of maximum energy which in the ocean extends to a depth of about 500 fathoms, or half-a-mile (Plates 6 to 20), and in the atmosphere, to a depth of 3000 fathoms, or three miles. Finally, the composition of both fluids is altered under the influence of solar heat; that of air through an increase in the quantity of moisture held in suspense, that of water by an increase in the percentage of salt held in solution.

THERMAL CIRCULATION.—It has been shown by Sir John Herschel that the unequal exposure of the different zones of the earth's surface to the rays of the sun must result in a system of atmospheric circulation composed of equatorial and polar currents; and by Lieutenant M. F. Maury, that the same inequality must be considered as the primary cause of a system of oceanic circulation also composed of polar and equatorial currents. The two distinguished philosophers have proved that these currents do not flow in the direction of the meridian, since, under the influence of the diurnal rotation of our planet from west to east, polar currents, as they move from a parallel of a lesser to one of a greater rotatory velocity, have a tendency to lag behind and to deviate in a westerly direction, while equatorial currents, in their progress from a lower to a higher latitude, have a tendency to deviate in the direction of the earth's rotation, *i.e.*, towards the east. All observations agree in establishing the fact that the internal economy of the atmosphere and of the ocean is regulated by such a system of circulation composed of equatorial and polar currents, and that the direction of these currents is affected by the earth's diurnal rotation in the manner above described.

VERTICAL AND HORIZONTAL EXTENSION OF THE TWO TERRESTRIAL ENVELOPES.—Before entering upon an examination of the phenomena of atmospheric and oceanic circulation, it is necessary to attach due importance to a condition sometimes overlooked in connection with occurrences which, in their *ensemble*, embrace immense areas of the surface of our planet—that is, the great disproportion which exists between the depth or vertical extension and the lateral or horizontal extension of the two terrestrial envelopes. The neglect of this condition is probably due to the exaggerated scale which it is necessary to adopt in graphical representations of oceanic and atmospheric sections, and to the difficulty of placing before our mental vision phenomena of such colossal proportions as we find realised in the great currents of the air and of the sea.

The average depth of the ocean, whether we estimate it at

Extension of the Two Terrestrial Envelopes. 49

two or three miles, is but a minute fraction of the length and breadth of an oceanic basin; so is the depth of the more active stratum of the atmosphere when compared with the areas of sea and land with which it is in contact.

A due consideration of this disproportion between horizontal and vertical extension leads to several conclusions of some importance to the student of the phenomena of oceanic and atmospheric circulation, namely :—1. That what in either system of currents has been called *horizontal* circulation must be the preponderating phenomenon, *vertical* circulation only occupying the second place. 2. That the original direction, volume, velocity, temperature, and composition of a current, considered as part of a system of thermal circulation, must undergo important modifications under the influence of the terrestrial areas with which the current comes in contact—an influence depending upon the distribution of land and water, the direction of the mountain ranges and coast lines, the configuration of the surface of the land and of the bottom of the sea, and other conditions present in a given area of the earth's surface. 3. That the currents of the ocean and of the atmosphere, while obeying their original tendency as thermal currents to flow in a certain direction—towards the Equator in the case of polar currents, towards the Poles in the case of equatorial currents —will ultimately move in the direction of least resistance. 4. That under the influence of local conditions, the general system of atmospheric or oceanic currents will resolve itself into as many different systems of circulation as there are distinct areas of land and water.

Sir John Herschel says, in his *Treatise on Astronomy* (sec. 197): "We have only to call to mind the comparative *thinness* of the coating which the atmosphere forms around the globe, and the immense *mass* of the latter, compared with the former (which it exceeds at least 100,000,000 times), to appreciate fully the *absolute command* of any extensive territory of the earth over the atmosphere immediately incumbent on it, in point of motion."

This remark, made with regard to the accelerating effect which the friction of the earth's surface exercises upon the rotatory velocity of the superincumbent atmosphere, applies with equal force to the powerful influence which the conformation of the surface of the solid earth's crust must exercise upon the atmospheric and oceanic currents with which this surface comes in contact.

Notwithstanding the daily accumulating mass of observations made in every part of the world, dissatisfaction has been frequently expressed at our imperfect insight into the laws which govern meteorological phenomena. Perhaps we look in vain for a direct manifestation of those laws under conditions constantly tending to modify the form under which they are presented to us, and the observer must be content to discover the expression of a general law under the disguise of ever varying and often contradictory phenomena. Every terrestrial area, both on sea and land, has its own system of atmospheric and oceanic currents; as every part of a continent, every valley has its own climate, subject to the general laws which govern the circulation of currents and the distribution of climate over the whole of our planet.

PARALLELISM BETWEEN OCEANIC AND ATMOSPHERIC CURRENTS. —On account of the more uniform conditions which prevail over oceanic areas in comparison with continental areas, the phenomena of currents are less complicated in the former than in the latter, and can be studied to greater advantage.

If we consult a wind-chart, we find that the direction of the prevailing currents of air which flow over the surface of the ocean agrees with the direction of atmospheric currents, such as would arise from the unequal distribution of solar heat over the surface of the rotating globe. We have currents of cold air flowing from high into low latitudes in a westerly direction, and currents of warm air passing in an easterly direction from the Tropics into the temperate and the polar regions. The effect of these currents moving in opposite directions is seen in the creation of several belts or areas of calms, one near the Equator, one at each Pole, and one near the parallels of lat. 30°. In accordance with theory, the belts of calms are due to an encounter which takes place in about lat. 30° between equatorial and polar currents. The former, coming from the Equator in the character of upper-currents, are supposed to descend in that latitude to the surface of the ocean, and to continue their course towards the polar regions as under-currents, having acquired an easterly tendency owing to the gradually decreasing rotatory velocity of the areas over which they flow, until, finally, they are arrested as they approach the Poles by the friction of the earth's surface. The latter, coming from the polar regions as upper-currents, descend near the same latitude towards the surface of the ocean, and continue their course towards the Equator as under-currents, gradually losing their westerly tendency until, in the vicinity of the Equator, they commence to rotate with the earth's surface and are no longer felt as easterly winds, thus producing the Equatorial belt of calms.

A current moving from the Equator towards the Pole will not acquire a decided tendency towards the east until it reaches the parallel of lat. 30°, as the diameter of rotation decreases very slowly at first, its total decrease between lat. 30° and 45° being greater than that between the Equator and lat. 30° (Fig. 13). It will, therefore, not manifest itself as a strong easterly current until it crosses the 30th parallel. On the other hand, a current flowing from the Pole towards the Equator, while having from the outset a strong tendency to lag behind in a westerly direction, on account of the steady increase of the diameter of rotation between the Pole and the parallel of lat. 30°, will, after crossing that parallel, gradually lose that tendency, and will, within 10° of the Equator, have acquired the rotatory velocity of the earth's surface, and therefore cease to show itself as a westerly current.

It thus appears that the Equator, the parallels of lat. 30°, and the Poles, constitute what may be termed the *critical* latitudes in the system of atmospheric circulation, and for similar reasons also in the system of oceanic circulation.

This state of matters, according to which we find the surface of our planet, as regards its two fluid envelopes, divided into belts of calms and belts of currents symmetrically distributed on each side of the Equator, is subject to considerable modifications from various causes. The first and most important of these causes is the division of the surface of our planet into areas of land and water which, alternately stretching across the Equator from one hemisphere into the other, intersect the parallel belts of calms and of currents at right angles. We have here carried out on a large scale one of those simple expedients by which Nature, in strict obedience to her laws, creates that endless variety of contrasting phenomena, which the philosopher, the poet, the artist, never cease to behold with wonder, and which, while it is the source of all beauty, is, at the same time, a necessary condition to the existence of all life. The result, in the present case, is the creation of numerous areas of atmospheric and oceanic circulation corresponding with the different areas of land and of water distributed on each side of the Equator, and the subdivision of the belts of calms into distinct areas of calms, of which we find one in each of the oceanic basins, in the North and South Atlantic, in the North and South Pacific, and in the Indian Ocean. (Plate 4 A.)

A comparison of these areas of calms with a chart of

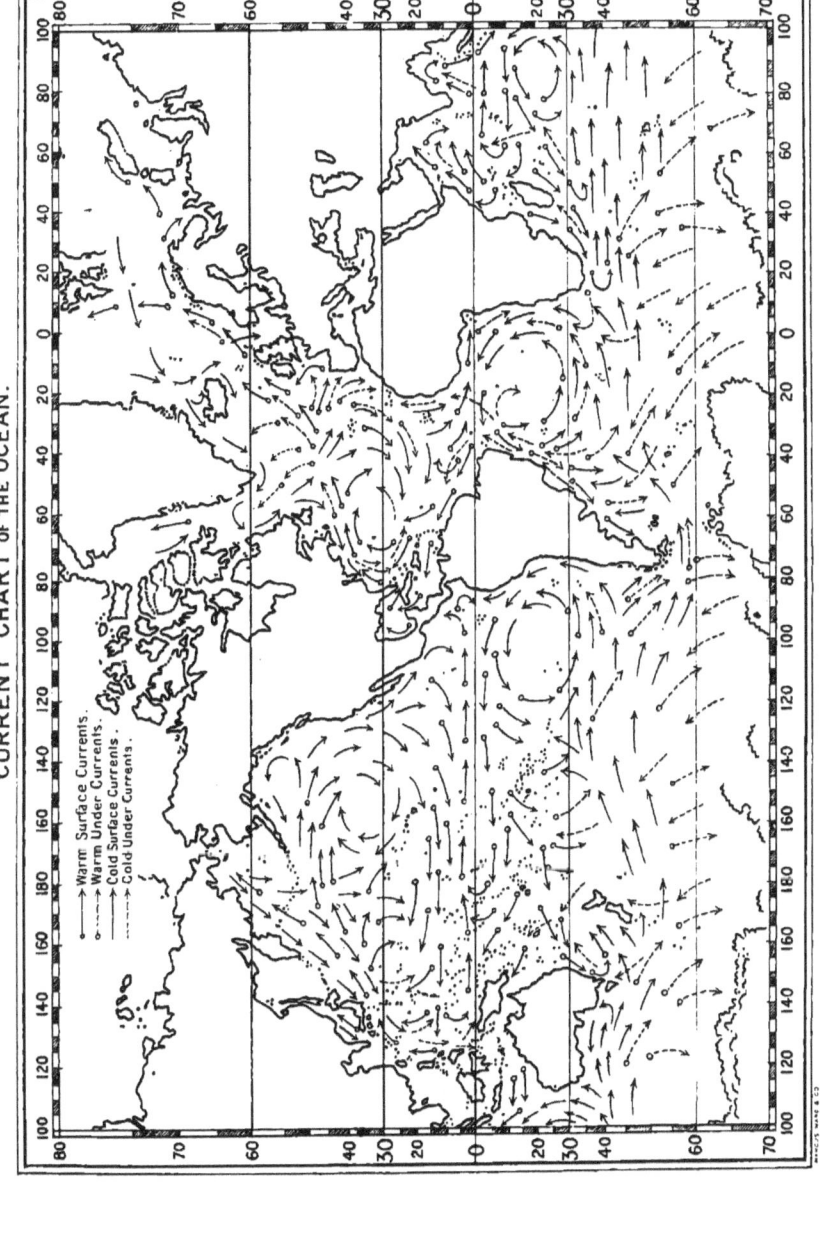

isobars shows that they form at the same time areas of high barometric pressure round which the atmospheric currents revolve—with the hands of the watch in the northern hemisphere, against the hands of the watch in the southern hemisphere, while the equatorial belt and the circumpolar regions form areas of low barometric pressure. If we now compare a chart of isobars and a wind chart with a chart of oceanic surface-currents, we find that there is a remarkable agreement between them. These currents are observed to revolve together with the winds round the areas of calms and of high barometric pressure placed near the centre of each oceanic basin, about the parallel of lat. 30°. The conclusion at which we arrive, that the winds are the cause of the surface-currents, seems obvious, although we must not forget that even in the absence of wind, the thermal circulation of the ocean would resolve itself into a system of surface and under-currents. It may be more in accordance with facts to suppose that atmospheric and oceanic currents mutually act and react upon each other. A current of water will either raise or lower the temperature of a stratum of air with which it remains in contact, and thus cause an inflow of air from neighbouring colder regions, or an outflow of air into adjoining warmer areas. A current of air will induce a surface-current in the stratum of water with which it comes in contact, or accelerate the velocity of a surface-current already existing, or change its direction, or arrest its motion. That the winds are a direct cause of oceanic surface-currents is a fact too well established by actual observation to require further proof. Perhaps the most striking example of their agency will be found in the complete reversal of the currents in the regions of the Monsoons.

There exists a second and hardly less important cause which tends to modify the general system of atmospheric and oceanic circulation. The sun, in its apparent progress from tropic to tropic, causes a change in the distribution of solar heat over the surface of our globe, by transferring the zone of maximum heat from one hemisphere to the other. As a necessary consequence, the volume, rate, and direction of the different currents are found to vary with the seasons; cold currents preponderating at one time of the year, and warm currents at another. At the same time the areas of calms and of high or low barometric pressure expand and contract, and are, to a limited extent, displaced.

SURFACE AND UNDER-CURRENTS.—It may be taken for granted that, water being a ponderable substance, and, as such, subject to the laws of gravity, the different strata of the ocean will be found arranged according to their weight, the heavier strata below, the lighter strata above. The weight of salt water varies with two conditions: temperature and percentage of salt held in solution. In the tropical belt, the water of the surface-stratum contains more salt, the increase being due to the evaporation caused by the rays of the sun. In the polar regions, the quantity of salt falls below the average on account of the greater proportion of fresh water derived from the melting of the ice and from precipitation.

The observations made on board the "Gazelle" have shown that there is a direct relation between the colour of sea water and the percentage of salt which it contains. The more salt it holds in solution the more intensely blue is its colour; the less salt it contains, the more greenish the colour is. In extra-tropical latitudes, we sometimes observe water of a beautiful blue colour, as, for example, in the Mediterranean and other nearly land-locked basins, where, the inflow of fresher water being more or less cut off, the percentage of salt is raised above the average by evaporation. We also observe it when crossing a current coming from tropical latitudes, such as the Gulf Stream. A green colour is sometimes met with in the tropics, in places where great rivers pour their masses of fresh water into the sea.

According to the extensive series of specific gravity observations on sea-water made on board H.M.S. "Challenger," by Mr. T. Y. Buchanan, M.A., chemist to the expedition, it appears that the specific gravity of salt water under the influence of temperature varies between a minimum of 1.021 and a maximum of 1.028 (to use round numbers), and the specific gravity, as affected by the percentage of salt contained in the water, between 1.024 and 1.027. If we select the sections surveyed by the "Challenger" between St. Paul Rocks at the Equator, the Cape of Good Hope, Kerguelen Land, and the Antarctic Circle, we find that the specific gravity of the surface water, according to the percentage of salt, commences with 1.027 near the Equator, falls to 1.026 towards lat. 40° S., to 1.025 between lat. 40° and 50° S., remains at that figure as far as lat. 60° S., and finally sinks to 1.024 in the immediate vicinity of the ice-barrier. On the other hand, the specific gravity of the surface-stratum, under the influence of temperature, commences with 1.023 near the Equator, rises to 1.026 towards lat. 40° S., attains 1.027 near lat. 50° S., and continues the same down to the ice-barrier. The temperature of the bottom-stratum in the different oceanic basins remains uniformly within a few degrees of 0° C., hence its specific gravity is generally about 1.028. The percentage of salt in the bottom-water was found to decrease from the Equator towards the polar regions, the specific gravity falling from 1.027 to 1.025.

It will thus be seen that the difference in specific gravity due to temperature is more than double the difference arising from the varying percentage of salt; whence we conclude that the order of the oceanic strata depends, in the first instance, upon temperature, in the second, upon the amount of salt held in solution.

An equatorial surface-current will remain such so long as its temperature is sufficiently high to render it lighter than the surrounding waters, but as during its progress towards higher latitudes its temperature decreases, it will, on account of its greater saltness, sink below the fresher waters of these latitudes, and continue its course as a warm under-current towards the polar regions. On the other hand, a polar surface-current, although composed of fresher water, will, on arriving at a certain latitude, sink below the tropical waters on account of its low temperature and consequent greater specific gravity, and continue its course towards the Equator as a cold under-current.

But the temperature of the ocean decreases not only from the Equator towards the Poles, but also from the surface towards the bottom. Hence, in the tropical regions, the warm but salt surface-water will sink on becoming cooled by its contact with the strata beneath, and impart some of its heat to the latter; while in the polar regions, the fresh but cold surface-water produced by the melting of the ice will, on becoming more salt by its admixture with the surrounding water, sink in its turn and lower the temperature of the strata with which it comes in contact.

The final result of these exchanges of temperature, which constitute what may be called the vertical circulation of the oceanic waters, appears in the oblique position, and the consequent spreading out of the isotherms as we recede from the Equator (Plates 9 and 19). The isotherm of 5° C., for example, which near the Equator is found at a depth of 300 fathoms, is met with at 600 fathoms in lat. 50° S. in the Southern Ocean, and at 800 fathoms in lat. 50° N. in the North Atlantic.

The Southern Ocean is the main feeder of its three gigantic offshoots—the Atlantic, the Pacific, and the Indian Oceans, which it supplies through the medium of both surface and undercurrents. The former, driven by the westerly winds against

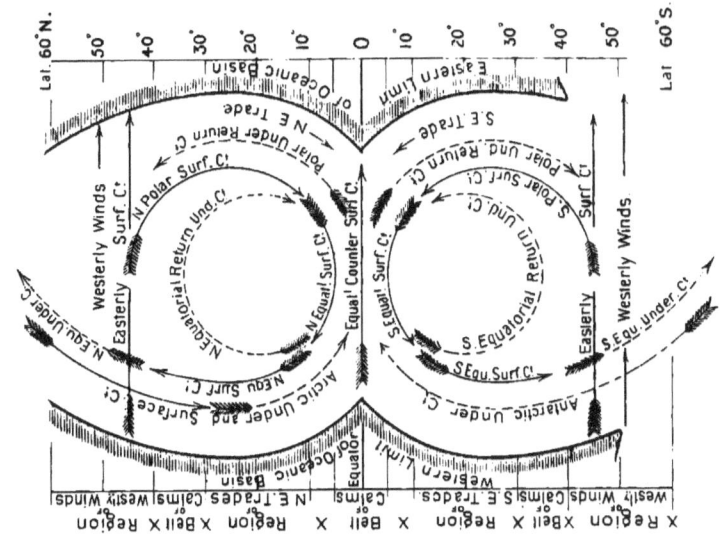

Fig.12.
General Diagram of Oceanic Circulation

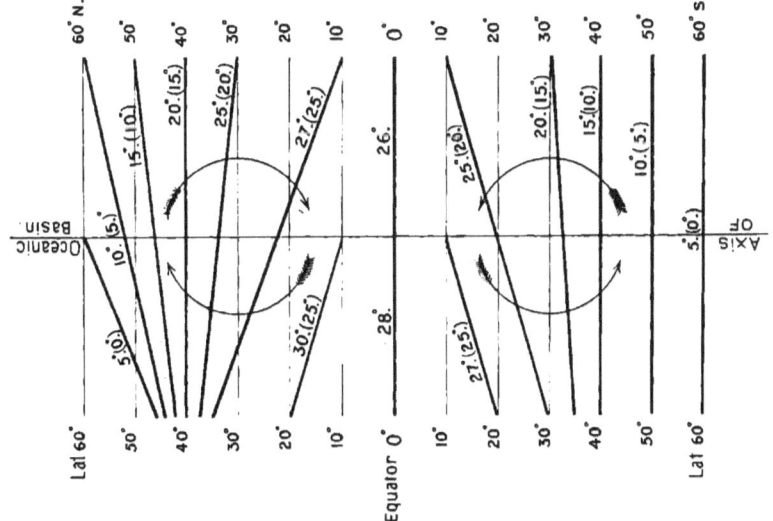

Fig.II.
Diagram of Oceanic Isotherms.

the west coast of Africa, Australia, and South America, are diverted northwards towards the Equator; the latter, piled up by the rotating earth against the east coasts of these continents, flow as under-currents in the same direction, both returning in the character of warm currents towards their old home at the Pole.

The annexed diagram represents the surface and under-currents which, in accordance with the above theoretical deductions, compose the system of circulation in our principal oceanic basins (Fig. 12).

CHAPTER IV.

THE TEMPERATURE SECTIONS SURVEYED BY H.M.S. "CHALLENGER" IN THE ATLANTIC.

From Teneriffe to Sombrero and St. Thomas—From St. Thomas to Halifax—Between Cape May, U.S., and Madeira—From Madeira to Tristan d'Acunha—Between Cape Palmas and Cape S. Roque—Between Cape S. Roque and Tristan d'Acunha—From the Falkland Islands to the Cape of Good Hope.

THE OCEANIC TEMPERATURE SECTIONS SURVEYED BY H.M.S. "CHALLENGER."—The accompanying diagrams and tables, especially constructed by the author for this essay, embody the principal results of the sounding operations carried on by the officers of H.M.S. "Challenger" during her cruise round the world between December, 1872, and May, 1876.

The isotherms of $2°.5$, $5°$, $10°$, $15°$, $20°$, and $25°$ C., have been selected, partly as affording a sufficiently correct representation of the distribution of temperature in the different oceanic sections which have been explored, partly because the above degrees of the Centigrade scale correspond with even numbers of the Fahrenheit scale, namely, $36°.5$, $41°$, $50°$, $59°$, $68°$, and $77°$. The intervening isotherms are, as a general rule, symmetrically arranged between these limits. As the temperature of $10°$ C. ($50°$ F.) fairly marks the point which divides what may be called *warm* water from *cold* water, the strata of a temperature above $10°$ C. have been coloured *red*, those below that temperature *blue*. The strata coloured *violet* are those in which the temperature of the water has been found to remain unchanged down to the bottom (Plates 15 and 16). The *yellow* or *buff* tint indicates where bottom has been reached within 1500 fathoms from the surface. It should also be observed that the scale of depth is greatly in excess of the scale of distance marked in degrees of latitude and longitude. For example, in Plate 9, a division of 100 fathoms is equal to two divisions, or two degrees of the horizontal scale, representing 120 nautical miles, or about 120,000 fathoms, so that the proportion between the two scales is as 1 to 1200—in other words, the depths in that diagram are 1200 larger than the distances indicated by the horizontal scale. The scale of depth, which stops at 1500 fathoms, represents only about three-fourths, and often only one-half, of the total depth of the oceanic basin; and from the lowest isotherm of $2°.5$ C., the temperature in most cases slowly decreases down to the bottom, the depth and temperature of which at each station are given in the table annexed to each diagram.

The station numbers are the same as those on the labels attached to the natural history specimens brought up by the dredge, the trawl, or the towing-net at each station. No doubt these specimens, of which there are more than one hundred thousand, embracing several hundreds of forms of animal life never before beheld by the eye of man, and therefore highly interesting not only to the student of zoology but to the public in general, will be permanently exhibited in the shape of a "Challenger Museum." Collected, as they have been, at a great sacrifice of time, money, and, sad to say, of life, including the ever-to-be-regretted death of Dr. Rudolf von Willemoes-Suhm, the promising zoologist attached to the scientific staff of H.M.S. "Challenger," they will compose a lasting monument of the generosity of the English nation, always ready to promote the cause of knowledge, and prove of more enduring interest to future generations than all the trophies of war bought at the price of general ruin.

SECTION FROM TENERIFFE TO SOMBRERO (Plate 6, Table I.).—This section stretches across the Atlantic in a west-south-westerly direction, and crosses the parallel of lat. 20° N. near

TABLE I.—TEMPERATURES OBSERVED BETWEEN TENERIFFE & SOMBRERO—*Feb, Mar, 1873.*

STATION NO.		1	2	4	5	8	10	13	15	17	18	20	22
LATITUDE AND LONGITUDE.		27° 24' N. 16° 55' W.	25° 52' N. 19° 14' W.	25° 28' N. 20° 22' W.	24° 20' N. 24° 28' W.	23° 12' N. 32° 56' W.	23° 10' N. 38° 42' W.	21° 38' N. 44° 39' W.	20° 49' N. 48° 45' W.	20° 7' N. 52° 32' W.	19° 41' N. 55° 13' W.	18° 56' N. 59° 35' W.	18° 40' N. 62° 56' W.
Surface Temp.	C. F.	18°.0	19°.5	19°.5	20°.0	19°.5	22°.2	22°.2	22°.5	23°.3	23°.3	23°.9	24°.4
ISOTHERM OF	25° 77°	—	—	—	—	—	—	—	—	—	—	—	—
	20° 68°	—	—	—	—	—	—	—	—	—	—	—	—
	15° 59°	—	—	—	0	—	80	100	85	120	110	100	110
	10° 50°	140	141	161	165	200	180	210	200	225	210	210	230
	5° 41°	350	335	340	355	365	350	380	350	350	360	290	360
	2°.6 36°.5	800	840	840	840	775	660	630	600	700	600	600	700
		—	—	1600	1600	—	1600	1300	1500	1500	—	1500	—
Bottom Temp.		2°.0	2°.0	—	2°.0	2°.0	1°.9	1°.9	1°.7	1°.9	1°.6	—	3°.0
Depth in Fms.		1890	1945	2220	2740	2800	2720	1900	2325	2385	2675	2975	1420

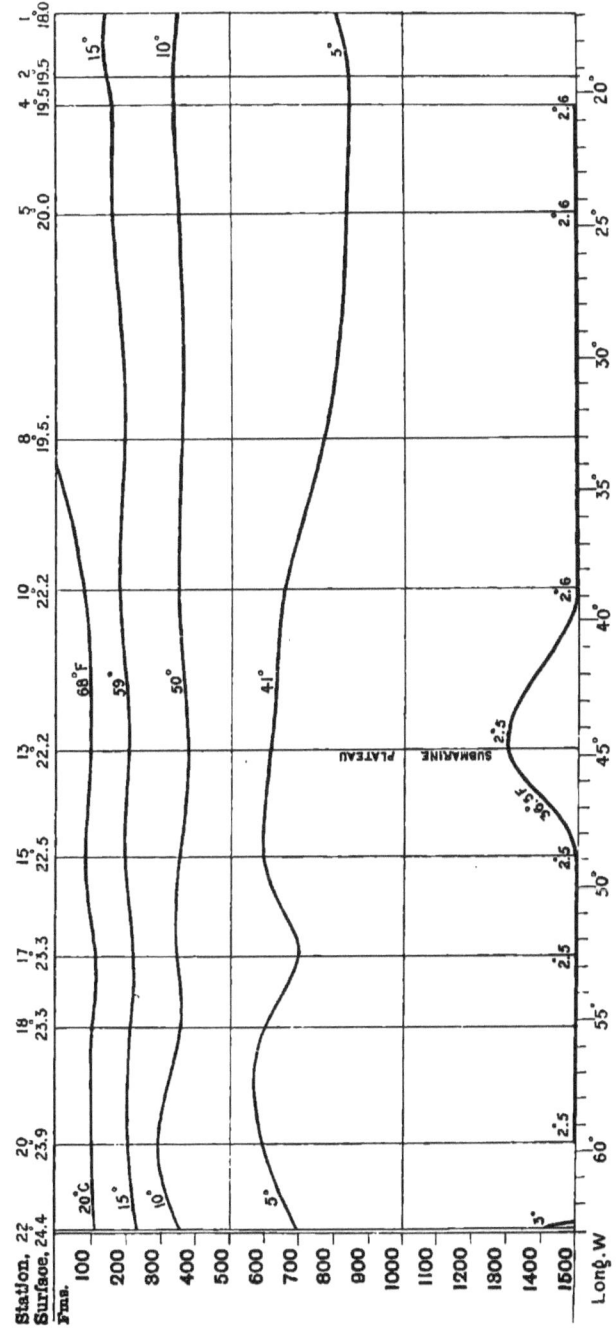

the Antilles (Plate 2). It affords an instructive example of the contrast which has been observed between the two portions of the North Atlantic divided from each other by the central plateau, as regards distribution of temperature. In the eastern basin the temperatures are lower at the surface, higher in the deeper strata than in the western basin; while in the latter they are higher at the surface, and lower in the deeper strata when compared with the former. The isotherm of 10° C., which throughout the section remains at about the same level, marks the turning-point of the change. Station 13, placed upon the central plateau, may be considered as dividing the two areas of circulation, which, however, as might be expected, encroach upon each other. In the western basin, the warm surface-stratum due to the *North Atlantic Equatorial Current*, and extending down to 100 fathoms, stretches eastwards beyond Station 13 as far as Station 10, and, gradually thinning off, disappears near Station 8, where it makes room for the *North Atlantic Polar Current*, which forms the surface-stratum of the eastern basin. The cold stratum below 400 fathoms, which in the west has a temperature of 5° C. at about 600 fathoms, shows in the east a fall of this isotherm down to 840 fathoms—in other words, a rise in the temperature of the lower strata, which, as the difference in bottom-temperature (Table I.) indicates, extends to the bottom. As we pass from the cold water accumulated to westward of the central ridge by the *North Atlantic Polar Under-current*, we enter already, at Station 15, into a warmer stratum caused by the mixture of the North Atlantic Polar Current with the North Atlantic Equatorial *Return Current*, flowing down together in the eastern basin. In the western basin, the surface and the deeper strata flow in opposite directions—one north, the other south, and the curves show a more or less abrupt transition from the warm upper strata to the cold lower strata; whilst in the eastern basin, the currents flowing in the same direction, *i.e.*, south, the heavier equatorial return current sinks through the lighter polar current, and the curves present a slow and gradual decrease of temperature from the surface to the bottom (compare Curve B of Station 5, Fig. 4, with the Equatorial Curve, Fig. 9, of Station 110, on the western slope of the central plateau).

At Station 13 may be observed a remarkable phenomenon, frequently noticed during the progress of the "Challenger" expedition, namely, *the simultaneous rise of the isotherms with the sea-bottom.* This phenomenon first attracted the attention of the officers of the U.S. Coast Survey as they were engaged in tracing the course of the Labrador current along the coast of the United States. This current was found to rise and fall with the sea-bottom over which it flows, and finally to force its way into the Strait of Florida at the high level of less than 300 fathoms from the surface, immediately below and in a direction contrary to the Gulf Stream current.

This circumstance seems to indicate that the great thermal currents which, without ceasing, tend to restore the oceanic equilibrium disturbed by the unequal distribution of solar heat, force their way from north to south, and from south to north, against every obstacle to their progress arising from the irregular conformation of the sea-bottom, and from the direction of the coast-lines which cross their path. They rise and fall with the sea-bottom, and accumulate their waters against the shores of islands and continents which stand in their way. There are also indications sufficient to show that the presence of land or submerged areas of elevation is not indispensable to the production of this phenomenon, and that currents flowing side by side but in different directions accumulate their waters against each other, in consequence of which the weaker current gives way to the stronger, and the waters of the lighter current flow over the surface of the heavier current, as is seen in the case of the Gulf Stream, which, along the United States coast, flows over

From St. Thomas to Halifax. 63

the Labrador current as the latter forces its way southward between the former and the coast of America.

The two depressions in the isotherms of 10° C. and 5° C., noticed in the western half of the section between Teneriffe and Sombrero, may thus be accounted for—at Station 22 by an accumulation of the water of the equatorial current against the inclined base of the Caribbee Islands, and at Stations 18 and 17 by the same current forcing its way between the masses of the polar under-current moving in an opposite direction. The *undulating form* of these isotherms shows that in the western half of the section there are two currents moving in different directions and contending against each other—a form which, at the surface of the ocean, assumes the character of alternating streaks of warm and cold water flowing in opposite directions. This phenomenon is realised on a large scale by the Gulf Stream current, which, in its progress northward, is split up into several branches by the Labrador current, and, as the latter suffers the same fate at the hand of the former, the scene of the contest is covered with alternate streaks of warm and cold water, the one flowing north, the other south. The Agulhas current, off the Cape of Good Hope, and the Kuro-Siwo stream, off the coast of Japan, furnish illustrations of the same phenomenon on a scale not much inferior, and it will occur wherever currents of different origin, and therefore of different temperature, weight, and chemical composition, meet each other.

For similar reasons, one current may present as solid an obstacle to the progress of another current as if it were a barrier of rock, and compel the latter either to alter its direction or to flow above or below the former—a phenomenon, as will be seen in the course of the following pages, also exhibited on a large scale in the system of oceanic circulation.

SECTION FROM ST. THOMAS TO HALIFAX (Plate 7, Table II.). —This section extends in a direction nearly due north along

TABLE II.—TEMPERATURES OBSERVED BETWEEN ST. THOMAS & HALIFAX—*March, May, 1873.*

STATION NO.	25	27	28	29	57	55	54	53	52	51	50
LATITUDE AND LONGITUDE.	19° 41' N. 65° 7' W.	22° 49' N. 65° 19' W.	24° 39' N. 65° 25' W.	27° 49' N. 64° 59' W.	Off Bermudas.	33° 20' N. 64° 37' W.	34° 51' N. 63° 59' W.	36° 30' N. 63° 40' W.	39° 44' N. 63° 22' W.	41° 19' N. 63° 12' W.	42° 8' N. 63° 39' W.
Surface Temp. C.	24°.4	24°.2	23°.9	22°.2	22°.8	21°.4	21°.4	22°.5	19°.6	15°.0	8°.0
ISOTHERM OF 25° / 77° F.	—	—	—	—	—	—	—	—	—	—	—
ISOTHERM OF 20° / 68° F.	110	140	80	65	40	20	20	40	—	—	—
ISOTHERM OF 15° / 59° F.	240	260	285	285	330	360	360	320	330	0	—
ISOTHERM OF 10° / 50° F.	350	390	425	420	455	460	—	450	450	160	—
ISOTHERM OF 5° / 41° F.	600	630	650	625	600	600	—	560	600	320	200
ISOTHERM OF 2°.5 / 36°.5 F.	1500	1450	1550	1450	1200	—	—	1550	1600	1400	—
Bottom Temp.	—	1°.5	1°.7	1°.6	—	—	—	1°.8	1°.5	1°.5	2°.8
Depth in Fms.	3875	2960	2850	2700	1575	2500	2650	2650	2800	2020	1250

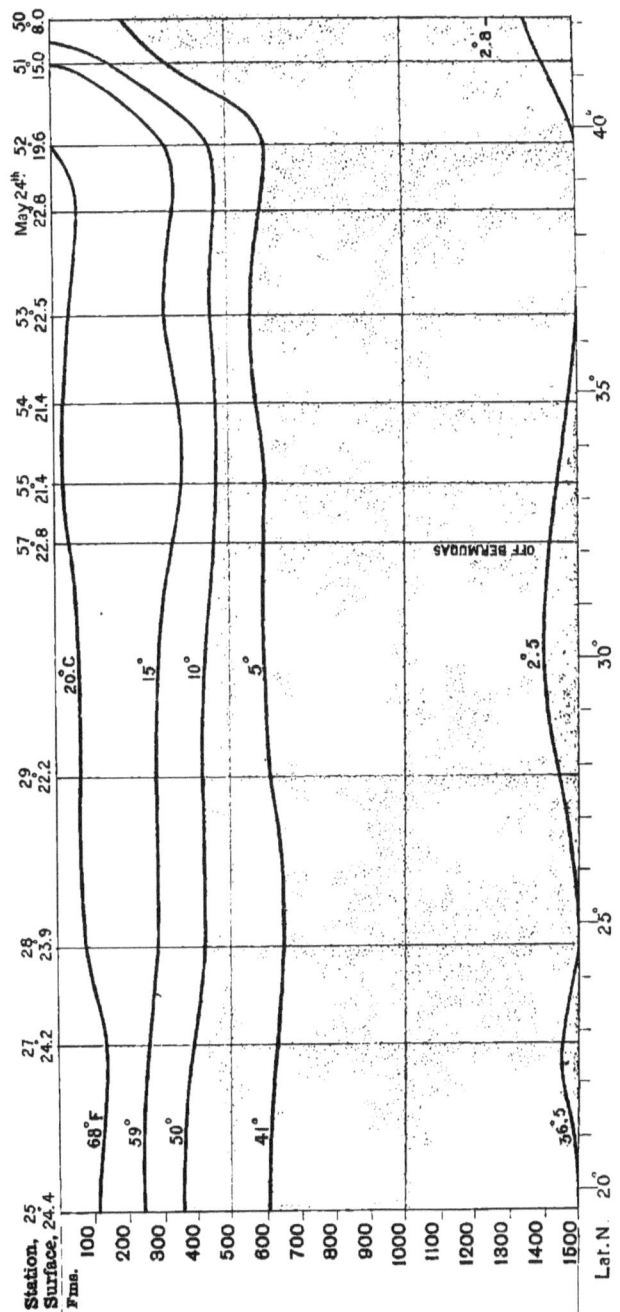

the meridian of long. 65° W., from St. Thomas, in the West Indies, to Halifax in Nova Scotia, the group of the Bermudas dividing the section into two nearly equal halves (Plate 2). An examination of the isotherm of 20° C., as well as of the surface-temperatures, shows that, between Station 27 and Station 28, we cross the northern limit of that portion of the North Atlantic Equatorial Current which flows outside the West Indian Islands. Reduced in depth, the warm surface-stratum continues towards the Bermudas, beyond which group it suffers further reduction by coming in contact with the Labrador current. At the station of the 24th May, we once more fall in with the equatorial current, namely, that portion of it which, after entering the Caribbean Sea and after making the circuit of the Gulf of Mexico, issues out of the latter through the Strait of Florida and flows along the U.S. coast under the name of the Gulf Stream. At the above station, the "Challenger" found a surface-stratum 50 fathoms thick of a nearly uniform temperature of 22°.8 C., only 1°.6 C. below the surface-temperature of the current outside the West Indies. At Station 52 we encounter the "cold wall" of the Labrador current, against which the Gulf Stream banks itself up during the whole of its course along the American coast; and at Station 51 and Station 50 we observe the rapid fall of temperature due to this cold current.

Those who have effaced the Gulf Stream off the Banks of Newfoundland, and have attributed to the North Atlantic Equatorial Current, or "North Atlantic drift-current" as it has been called, the vast masses of warm water which occupy the basin of the North Atlantic as far north as Spitzbergen and Baffin Bay, and those who supported the opposite theory giving all the credit to the Gulf Stream, were probably both partly right and partly wrong in their conclusions. The "Challenger" observations leave little doubt but that the Gulf Stream is a branch of the equatorial current, separated from the latter during its course through the Caribbean Sea and the Gulf of Mexico, and joining it again after coming out of Florida Strait, from which place it forms the western edge of the great mass of equatorial waters during its further progress towards the north.

The Channel of Yucatan, by which the Equatorial Current enters the Gulf of Mexico, presents a much wider section than the Florida Channel, whence the current issues under the name of the Gulf Stream, and it seems as though more water flowed into the gulf than out of it, unless it flow out with increased velocity. It has been calculated that a difference of level amounting to two feet—a difference which falls within the error of even the most careful survey embracing so large an area—creates sufficient pressure to force the water through the Strait of Florida at the rate of four miles an hour, at which the Gulf Stream is known to flow out of the Strait. A similar phenomenon, occurring under similar conditions, may be observed in connection with the Kuro-Siwo current. A branch of the North Pacific Equatorial Current flows into the basin situated between the Philippine and the Ladrone Islands, which basin, like the Caribbean Sea, is separated from the ocean by a chain of islands, the projecting points of a submarine ridge, and the northern and narrow half of this basin stands in the same relation to the southern half as the Mexican Gulf to the Caribbean Sea. The current, after passing along the east coast of the Philippines, of Formosa, and of the islands which connect the latter with Japan, has to force its way, and, like the Gulf Stream, in the face of a contending polar current, over the shallow barrier which joins Japan to the chain of islands terminating with the Ladrone Group. After crossing this barrier, it unites itself to the portion of the North Pacific Equatorial Current which flows along the eastern side of these islands, and the two combined form the Kuro-Siwo current, whose waters are traced

From St. Thomas to Halifax.

through Behring Strait into the Arctic basin, and eastward as far as the west coast of North America.

The isotherms of 15° C. and 10° C., in the section between St. Thomas and Halifax, continue to descend to a lower level as the temperature of the intermediate strata increases with the distance from the equator, until, at Station 52, we enter the polar current. The portion of the Atlantic between Halifax and Bermudas is occupied by alternate streaks of warm and cold water, as will appear from the following observations made on board H.M.S. "Challenger."

After leaving Halifax on the 19th of May, the surface-temperature marked a steady increase from 4° C., to 10° C., when, between 3 and 7 a.m. of the 22nd May, a rapid rise of the temperature betrayed the existence of a belt of warmer water. The latter attained a temperature of 17° C. between 5 and 7 p.m. of the same day, but at midnight it fell to 12°.2 C., to rise half-an-hour afterwards, at 12.30 a.m. of the 23rd, to 15°.2 C. Between that hour until the arrival of the ship near Bermudas several alternate streaks of warm and cold water were passed through, the former of a temperature from 22° to 23° C., the latter from 18° to 20° C. It will be observed that the water of the Gulf Stream Current was only cooled down to the extent of 1° C. during its passage from the section between Bermudas and Sandy Hook to the section between Halifax and Bermudas.

The centre of the first warm belt was reached at 8.30 a.m. of the 23rd May, that of the second at 1 a.m. of the 24th, of the third at 8 a.m., of the fourth at midnight of the same day, and of the fifth at 1.30 p.m. of the 26th, the whole of the 25th May having been occupied in traversing a broad belt of colder water. In the vicinity of Bermudas the surface-temperature once more rose to 23° C.

TABLE OF SURFACE-TEMPERATURES BETWEEN HALIFAX AND BERMUDAS.

Date, 1873.	Station.	Latitude at Noon.	Hour.	Surface Temperature.	Observations.
May 9	Halifax.	44° 39' N.	—	2°.2 C.	
,, 19	,,	,,	—	4°.0 C.	Cold streak.
,, 20	49	43° 3' N.	—	5°.0 C.	
,, 21	50	42° 8' N.	—	8°.0 C.	
,, 22	51	41° 19' N.	3 a.m.	10°.0 C.	
,, ,,	,,	,,	7 ,,	14°.2 C.	
,, ,,	,,	,,	Noon	15°.2 C.	
,, ,,	,,	,,	5 to 7 p.m.	17°.0 C.	Warm streak.
,, ,,	,,	,,	10 p.m.	15°.3 C.	
,, ,,	,,	,,	Midnight	12°.2 C.	Cold streak.
,, 23	52	39° 44' N.	12.30 a.m.	15°.2 C.	
,, ,,	,,	,,	1 ,,	18°.2 C.	
,, ,,	,,	,,	1.30 ,,	20°.0 C.	
,, ,,	,,	,,	4 ,,	21°.6 C.	
,, ,,	,,	,,	8.30 ,,	22°.0 C.	Warm streak.
,, ,,	,,	,,	9 ,,	19°.3 C.	Cold streak.
,, ,,	,,	,,	11.30 p.m.	20°.0 C.	
,, ,,	,,	,,	Midnight	21°.4 C.	
,, 24	—	38° 32' N.	1 a.m.	22°.0 C.	Warm streak.
,, ,,	,,	,,	3 ,,	21°.8 C.	
,, ,,	,,	,,	3.30 ,,	20°.0 C.	
,, ,,	,,	,,	5 ,,	19°.4 C.	Cold streak.
,, ,,	,,	,,	5.30 ,,	21°.1 C.	
,, ,,	,,	,,	8 ,,	23°.1 C.	
,, ,,	,,	,,	8.30 ,,	22°.8 C.	Warm streak.
,, ,,	,,	,,	6 p.m.	22°.8 C.	
,, ,,	,,	,,	8 ,,	21°.1 C.	
,, ,,	,,	,,	9 ,,	19°.4 C.	Cold streak.
,, ,,	,,	,,	10 ,,	21°.7 C.	
,, ,,	,,	,,	10.30 ,,	22°.2 C.	
,, ,,	,,	,,	Midnight	22°.2 C.	Warm streak.
,, 25	—	37° 7' N.	1 a.m.	22°.2 C.	
,, ,,	,,	,,	1.30 ,,	20°.0 C.	
,, ,,	,,	,,	6.30 ,,	18°.0 C.	
,, ,,	,,	,,	7 ,,	18°.0 C.	Cold streak.
,, ,,	,,	,,	The rest of the day	19°.0 C. to 20°.9 C.	
,, 26	53	36° 30' N.	3 a.m.	19°.8 C.	
,, ,,	,,	,,	4 ,,	21°.4 C.	
,, ,,	,,	,,	Noon	22°.8 C.	
,, ,,	,,	,,	1.30 p.m.	23°.1 C.	
,, ,,	,,	,,	4 ,,	23°.1 C.	Warm streak.
,, ,,	,,	,,	6.30 ,,	22°.8 C.	
,, ,,	,,	,,	8 ,,	22°.8 C.	
,, ,,	,,	,,	Midnight	22°.2 C.	
,, 27	54	34° 51' N.	12.30 a.m.	22°.5 C.	
,, ,,	,,	,,	7 p.m.	22°.2 C.	
,, ,,	,,	,,	8 30 ,,	21°.1 C.	
,, 28	55	33° 20' N.	1.30 a.m.	20°.5 C.	Cold streak.
,, ,,	,,	,,	7.30 ,,	20°.8 C.	
,, ,,	,,	,,	8 ,,	21°.2 C.	
,, ,,	,,	,,	Midnight	22°.2 C.	
,, 29	56	} Off Bermudas.	—	22°.2 C. to 23°.0 C.	Warm streak.
,, 30	57				
,, 31	Bermudas.	32° 15' N.	—		

The isotherms of 5° C. and 2°.5 C. retain an almost uniform level, the former at 600 fathoms, the latter at 1500 fathoms, as far as Station 52, where they both commence to rise rapidly and come to the surface off Halifax, as seen in the above table. The isotherm of 2°.5 C. rises with the sea-bottom around Bermudas. The bottom-temperature of 1°.5 C. is found at Station 51 at 2920 fathoms, at Station 52 at 2800 fathoms, and at Station 27 at 2960 fathoms.

SECTION BETWEEN CAPE MAY AND MADEIRA (Plate 8, Table III.).—This section crosses the North Atlantic from west to east between the latitudes of 32° and 38° N. It presents at Station 45, distant 140 miles from Sandy Hook, a section of the Labrador current, and at Station 43 a section of the Gulf Stream current. The latter formed, at the time of the "Challenger's" visit, the 1st of May, 1873, a surface-current about 60 miles broad and 100 fathoms deep, flowing in an east-northeasterly direction at the rate of three miles an hour. Later in the season the volume and velocity of this great oceanic river are much greater. As seen in the diagram, the Gulf Stream constitutes the extreme western border of the enormous mass of water which composes the North Atlantic Equatorial Current, and virtually flows over the Labrador Current found immediately below it. The temperature of the current was 23°.9 C. at the surface, 20° C. at 80 fathoms, and 18°.3 C. at 100 fathoms. But at 125 fathoms it had fallen already to 13°.8 C., and at 350 fathoms to 7°.1 C. At Station 44, just beyond the edge of the Gulf Stream, the temperature at the surface was 11°.1 C., at 125 fathoms 8°.3 C., and at 350 fathoms 3°.9 C. The distance between Station 43 and Station 44 is little over 60 miles.

The table on page 71 shows the remarkable changes of temperature observed by the "Challenger" in crossing the Gulf Stream.

TABLE III.—TEMPERATURES OBSERVED BETWEEN CAPE MAY, U.S., BERMUDAS, AZORES, AND MADEIRA—*April to July, 1873.*

Station No.	Latitude and Longitude	Surface Temp. C. / F.	25° / 77°	20° / 68°	15° / 59°	10° / 50°	5° / 41°	2°.6 / 36°.5	Bottom Temp.	Depth in Fms.
			ISOTHERM OF							
45	34° 10′ N, 72° 10′ W	10°.0	—	—	—	0	270	1200	2°.4 1°.7	1240
44	37° 25′ N, 71° 40′ W	11°.1	—	—	—	95	220	1200	1°.7	1700
43	36° 23′ N, 71° 51′ W	23°.9	—	80	150	225	650	—	—	2600
42	35° 58′ N, 70° 39′ W	18°.3	—	—	—	—	830	1650	1°.8	2425
41	36° 5′ N, 69° 54′ W	18°.3	—	—	350	470	615	—	—	—
39	34° 3′ N, 67° 32′ W	—	—	360	475	620	—	—	—	2850
38	33° 3′ N, 66° 32′ W	21°.1	—	40	385	500	665	1650	1°.8 1°.8	2600
37	32° 19′ N, 65° 39′ W (Off Bermudas)	20°.0	—	0	380	500	700	1550	1°.7	2650
57	Off Bermudas	22°.8	—	40	330	455	600	1200	—	1575
59	32° 54′ N, 63° 22′ W	23°.3	—	40	335	450	615	1400	1°.7	2360
60	34° 18′ N, 58° 56′ W	22°.0	—	25	300	425	600	—	1°.5	2575
61	34° 54′ N, 56° 38′ W	21°.7	—	20	320	450	—	—	1°.5 1°.8	2850
62	35° 7′ N, 55° 32′ W	21°.1	—	15	275	430	600	1600	—	2875
64	35° 35′ N, 50° 27′ W	23°.9	—	30	295	430	600	—	—	—
65	36° 33′ N, 47° 58′ W	22°.5	25	350	530	750	1650	—	1°.7	2700
66	37° 24′ N, 44° 14′ W	21°.1	—	20	335	500	690	—	1°.8	2750
67	37° 54′ N, 41° 44′ W	21°.1	—	40	335	460	620	1600	1°.8	2700
69	38° 23′ N, 37° 21′ W	21°.7	—	20	215	440	700	1550	1°.7	2200
71	38° 18′ N, 34° 48′ W	21°.7	—	15	90	310	600	1400	2°.2	1675
72	38° 34′ N, 32° 47′ W	21°.1	—	15	150	340	690	1240	2°.8	1240
73	38° 36′ N, 31° 14′ W	20°.6	—	25	120	380	740	—	3°.7	1000
76	38° 11′ N, 27° 6′ W	21°.1	—	20	60	365	780	—	4°.2	900
78	37° 24′ N, 25° 13′ W	21°.7	—	20	60	340	—	—	—	1000
79	36° 21′ N, 23° 31′ W	22°.0	—	20	60	360	840	1350	1°.6	2025
80	35° 3′ N, 25° 25′ W	21°.7	—	15	55	455	850	—	1°.8 1°.8	2660
82	33° 46′ N, 17° 61′ W	21°.5	—	15	80	440	880	1500	1°.8	2400

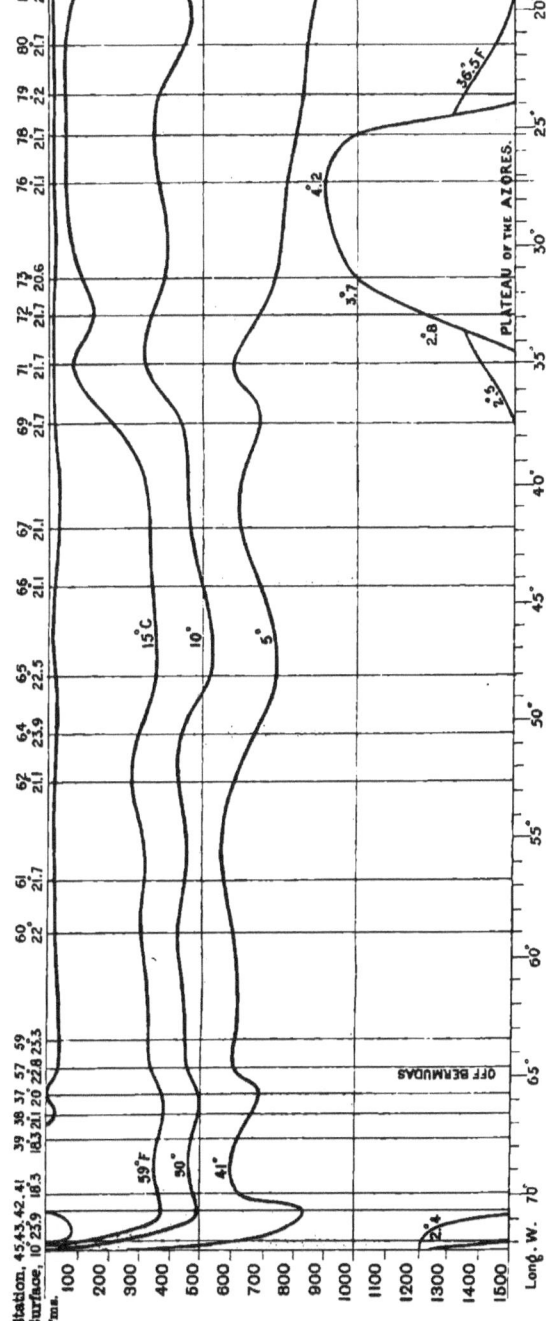

Date, 1873.	Station.	Latitude at Noon.	Longitude at Noon.	Hour.	Surface Temperature.
April 30th	42	35°58′	70°39′	1 p.m.	18°.6 C.
,,	,,	,,	,,	2 ,,	22.0
,,	,,	,,	,,	7 ,,	22.8
May 1st	43	36°23′	71°51′	6 a.m.	22.9
,,	,,	,,	,,	7 ,,	23.9
,,	,,	,,	,,	8 p.m.	23.0
,,	,,	,,	,,	9 ,,	20.6
,,	,,	,,	,,	11 ,,	19.4
,,	,,	,,	,,	Midnight	13.3

Passing along the section from west to east, we find inside the Gulf Stream an area of colder water extending from Station 41 to Station 39, then a streak of warmer water at Station 38, and one of colder water at Station 37, westward of Bermudas. These currents of colder water are probably a continuation of the streaks of low temperature met with between Halifax and the latter islands. At Station 57, off Bermudas, we enter an area of warm water of nearly the same temperature as the Gulf Stream, which can be traced as far as Station 60. Another area of the same temperature is met with further east at Station 64 and Station 65. The nearly uniform surface-temperature, varying from 21° to 22° C., found between these areas and in the rest of the section as far as Station 82, may be considered as the surface-temperature of this region of the North Atlantic in the months of June and July. The undulating level of the isotherms of 15°, 10°, and 5° C. betrays the struggle going on between contending currents, the warm water retaining the upper hand until we reach Station 71, on the western slope of the plateau of the Azores. This Station, like Station 10 on the eastern slope of the same plateau in the section between Teneriffe and Sombrero (Plate 6), fixes the limit between two areas totally different as regards the distribution of temperature, and a line drawn between the two stations will divide this portion of the Atlantic into an eastern area, in which the polar current predominates, and into a western area, where the equatorial current has the upper hand.

A glance at the chart will show that the Azores are placed in the direct continuation of the axis of Davis Strait, and there can be little doubt but that the sudden rise of the isotherms between Station 69 and Station 72 is due to a cold current, a branch of the Labrador current. The latter, as it arrives off the great bank of Newfoundland, spreads out like a fan—one branch, running in a south-westerly direction, forms the cold current of the coasts of Nova Scotia and the United States; the other branch, a south-easterly continuation of the original current, flows straight towards the Azores, to the westward of which islands it forms a powerful current more than a hundred miles broad. One part of this current probably runs down along the western slope of the plateau, while another part sweeps over the plateau in a south-easterly direction, carrying along with it a portion of the equatorial current, and thus composing that mixture of cold and warm water which we find in the channel between the Azores and Madeira, and which flows down outside the Canaries and the Cape de Verde Islands (Plates 8 and 9, Curves Figs. 3 and 4).

A line drawn between Newfoundland and the Azores passes over the area where the equatorial and the polar currents encounter each other, both coming out of the conflict profoundly altered in their constitution. To the northward of this line we find the equatorial current flowing in a north-easterly direction towards the Færoe Islands, sending off branches into Davis Strait and towards Iceland on one side, and towards the coast of Portugal, the Bay of Biscay, and the western shores of the British Islands on the other side. In this northern part of the North Atlantic the equatorial current generally assumes the

Between Cape May and Madeira. 73

character of an intermediate current, which extends from a depth of 100 fathoms down to 800 and 900 fathoms from the surface. The existence of this enormous mass of warm water, nearly a mile thick, off the west coasts of Europe, is amply proved by the temperature-soundings of H.M.S. "Porcupine" between the Færoe Islands and the Straits of Gibraltar. Its depth necessarily decreases as the current proceeds northward, but it still forms a layer 500 fathoms thick, of a temperature from 10° C. to 8° C. in the channel between Rockall and Scotland, and between the Hebrides and the Færoe Islands. It dwindles down to about 150 fathoms as it sweeps over the area of Arctic water (below freezing-point) between the Færoes and Scotland discovered by H.M.S. "Lightning," then follows the coast of Norway, and finally disappears in the seas of Spitzbergen and Novaya Zemlya. In about lat. 65° N., near the coast of Norway, it forms a surface-stratum about 200 fathoms thick of an average temperature of 7° C. In the same latitude, the zero-point is reached at a depth of 300 and 400 fathoms.

A comparison of the section between St. Thomas and Halifax with the section between Cape May and the Azores shows that the portion of the North Atlantic which they traverse forms an area of nearly equal distribution of temperature, and that this area coincides with what is known under the name of the *Sargasso Sea*. It extends from Station 28 to Station 52, or from lat. 25° to lat. 40° N. along the section between St. Thomas and Halifax, and it occupies the whole space between the edge of the Gulf Stream, near Station 41, and Station 69, near the Azores, or between long. 70° and long. 38° W. This area, about 1000 miles in diameter, may be fitly termed an immense whirlpool in the middle of the North Atlantic, whose waters revolve with the hands of the watch. It is covered with a surface-stratum of warm water about 300 fathoms thick,

F

in which the temperature slowly decreases from about 22° C. at the surface, to 16° C. at a depth of 300 fathoms. This stratum is represented by the protuberance or hump in Curve A, Fig. 2, and Curve A, Fig. 3. Throughout this area, the isotherm of 15° C. is found at an average depth of 350 fathoms, that of 10° C. at 450 fathoms, that of 5° C. at about 600 fathoms, and the isotherm of 2°.5 C. at 1500 fathoms, occasionally falling below 1600 fathoms.

SECTION FROM MADEIRA TO TRISTAN D'ACUNHA (Plate 9, Table IV.).—This section traverses the Atlantic Ocean from the parallel of lat. 30° N. to the parallel of lat. 37° S. along the meridian of long. 20° W., and may, as regards the distribution of temperature, be divided into three parts—a central belt, extending across the equator from lat. 15° S. to lat. 15° N., bounded on each side by the belts respectively belonging to the South and to the North Atlantic.

The most remarkable feature of the central belt is the rapid decrease of temperature in the surface-stratum of the ocean, amounting to from 15° to 19° C. within less than 200 fathoms, in comparison with the much slower decrease in the northern and southern belt, in the lower strata of which we observe a gradual increase of temperature as we recede from the equator, much more rapid, however, in the former than in the latter.

The maximum increase and decrease is seen in the diagram of the section to extend from Station 346, or lat. 3° S., to Station 95, or lat. 13° N., limits which coincide within a few degrees with the equatorial limits of the trade-winds, so that this belt may be considered as identical with the equatorial belt of calms. An examination of the section of the Pacific Ocean (Plate 19) shows that the same phenomenon has been observed in that ocean also. The increase is naturally due to the excess of solar heat in the equatorial belt, the decrease to the presence of a stratum of cold water within a few hundred fathoms from the surface. It appears from the diagram of the section between Cape Palmas and Cape S. Roque (Plate 10) that this stratum of cold water stretches right across the Atlantic, occupying the whole length of the equatorial belt, and that it is colder in the western half than in the eastern. The isotherm of 5° C. rises between Station 104 and Station 106 from 460 fathoms to 260 fathoms, its average level in the western portion towards Cape S. Roque being at a depth of 300 fathoms, and in the eastern portion towards Cape Palmas at 400 fathoms. On the other hand, the isotherms in the northern belt, from Station 95 to Station 84, show that the stratum between 100 and 500 fathoms is warmer than in the equatorial belt; while in the southern belt, between Station 339 and Station 342, the water between 250 and 500 fathoms is colder than in the equatorial belt. Between 500 fathoms and 1000 fathoms the same difference may be observed, the northern belt being warmer, the southern colder, than the central belt. The conclusion we arrive at seems obvious. As the cold stratum between 100 and 500 fathoms in the equatorial belt cannot come from the northern belt, which is warmer, it must come from the southern belt, which is colder at the same depths, and especially it must come from that portion of it which is situated between Station 339 and Station 342, or, more correctly speaking, from the belt which lies immediately to the southward of the equatorial belt, between lat. 10° S. and lat. 17° or 18° S. There must, therefore, be a current of cold water flowing from the southern belt into the equatorial belt above the level of 500 fathoms, and it is this current which forms the stratum of cold water found between 100 and 500 fathoms below the surface of the equatorial belt. According to the theory of thermal circulation, the superheated water of the equatorial region is constantly being transferred to higher latitudes, and is as constantly replaced by cold water pouring in from the polar regions. To

TABLE IV.—TEMPERATURES OBSERVED BETWEEN MADEIRA, ASCENSION, AND TRISTAN D'ACUNHA—*July, August, 1873; March, April, 1876.*

Station No.	135	335	336	337	338	339	340	341	342	343	345	346	347	348	102	101	100	99	98	97	96	95	92	91	90	89	88	85	84
Latitude and Longitude	37°20'S, 12°20'W	32°24'S, 13°5'W	27°54'S, 13°13'W	24°38'S, 13°36'W	21°15'S, 14°2'W	17°26'S, 13°52'W	14°33'S, 13°42'W	12°16'S, 13°44'W	9°43'S, 13°51'W	8°3'S, 14°27'W	5°45'S, 14°25'W	2°42'S, 14°41'W	0°15'S, 14°25'W	3°10'N, 14°51'W	3°8'N, 14°49'W	5°48'N, 14°20'W	7°1'N, 15°55'W	7°53'N, 12°26'W	9°21'N, 18°28'W	10°25'N, 20°30'W	12°51'N, 22°18'W	13°36'N, 22°49'W	17°54'N, 24°41'W	19°41'N, 24°9'W	20°58'N, 22°57'W	22°18'N, 22°2'W	23°35'N, 21°18'W	28°42'N, 18°9'W	30°38'N, 18°5'W
Surface Temp. C.	12°.0	23°.0	24°.4	25°.0	24°.7	24°.4	25°.1	26°.1	26°.7	27°.1	28°.2	28°.2	27°.8	28°.9	25°.6	26°.2	26°.1	25°.1	25°.6	25°.7	25°.6	25°.9	26°.1	23°.7	23°.3	23°.0	22°.0	20°.7	21°.7
Isotherm 25°/77°	—	—	—	0	—	—	0	30	40	30	35	20	20	25	15	10	15	40	20	25	15	7	—	20	30	40	20	5	25
20°/68°	—	35	35	60	70	55	70	60	60	45	55	30	45	35	45	30	35	80	30	50	30	15	50	50	150	160	140	100	140
15°/59°	35	105	100	150	125	100	95	90	75	65	70	35	75	55	100	75	60	190	150	160	150	35	130	290	300	310	310	340	400
10°/50°	65	265	220	255	200	185	160	160	160	140	135	175	180	185	210	175	190	500	500	440	500	200	235	640	660	750	755	—	—
5°/41°	360	430	380	380	330	265	350	360	475	390	465	370	480	400	360	380	440	—	—	1150	—	470	575	—	1250	1350	—	—	875
2°.5/36°.5	600	750	1200	1240	1400	1415	1500	—	—	—	1500	1460	1600	1400	1250	—	1400	—	—	—	—	1400	1400	—	—	—	—	—	—
Bottom Temp.	—	2°.3	1°.9	2°.5	1°.8	2°.5	2°.6	3°.0	2°.6	4°.5	2°.1	0°.4	1°.7	—	1°.7	1°.7	—	—	2°.0	1°.8	—	1°.8	—	1°.8	1°.8	1°.8	1°.7	—	—
Depth in Fms.	1000	1425	1890	1240	1990	1415	1500	1475	1445	425	2010	2350	2250	—	2450	2500	2425	—	1750	2575	—	2300	1975	2075	2400	2400	2300	1125	2400

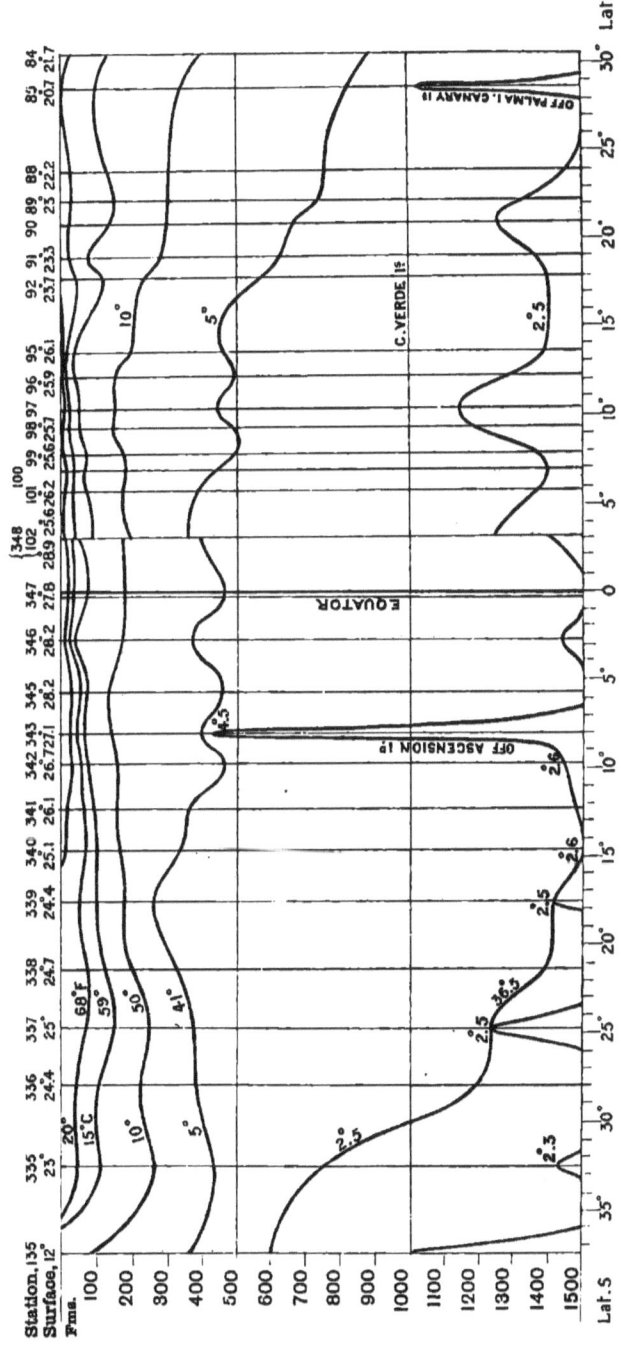

speak more definitely, the warm surface-water at the equator is replaced by water flowing in from the colder strata immediately adjoining, which must result in the formation and maintenance of a stratum of cold water immediately below the surface of the equatorial belt. This cold stratum, in the present case, is found to extend to a depth of about 400 fathoms.

In this sense the cold water may be said to rise up towards the surface at the equator, but if we compare the thickness of the strata under consideration with their horizontal extension in latitude and longitude, whatever movement may take place in a vertical direction must be absolutely insignificant in comparison with the horizontal currents thus created. This inflow of cold water will take place more or less along the whole boundary between the equatorial belt and the colder belt immediately adjoining, and this agrees with the extension of the cold stratum along the whole section between Cape Palmas and Cape S. Roque, the current being stronger in the western half than in the eastern. There are several indications of a large mass of cold water moving across the equator and the plateau of St. Paul Rocks in a north-easterly direction, and forming an under-current, which may be traced as far as the Cape de Verde Islands along the West Coast of Africa, and between the latter and the Canaries, Madeira, and even the Josephine Bank, off the Straits of Gibraltar.

The abrupt gradient of the temperature-curve which is so characteristic of the equatorial stations (Curve Fig. 9) is also observed at Station 95, Station 96, and Station 97, and to a slightly lesser extent at Stations 98, 99, and 100. In fact, this form of curve makes its appearance immediately south of the Cape de Verde Islands, and thence extends towards the equator. The cold current, after crossing the equator in an oblique direction from S.W. to N.E., probably divides itself off the coast of Africa into two branches, one flowing north, between the Cape de Verde Islands and the African coast, the other flowing south, past Sierra Leone towards Cape Palmas. At Stations 95, 96, and 97, in the direct line of this cold under-current, the decrease of temperature between the surface and 100 fathoms amounts to 15° C. The temperature observations made by the "Challenger" on the return journey at Stations 350, 351, 352, in the same area, off Sierra Leone, prove the same extraordinary decrease of temperature in the first hundred fathoms, amounting respectively to 15°.1, 14°.3, and 12°.6 C. The remarkable rise of the isotherm of 2°.5 C. in this area also deserves to be noticed. At Station 97, the temperature at 1500 fathoms was found to be 1°.7 C. Proceeding further north, at Stations 89 and 90, between the Cape de Verde Islands and the Canaries, the temperature at 1100 fathoms was only 3°.2 C. and 3° C. respectively. At a sounding taken on the 6th February, 1873, between Teneriffe and Madeira, the temperature at 1975 fathoms was ascertained to be 1°.6 C.; on February 6th, to the east of Madeira, 1°.8 at 2225 fathoms; and on January 30th of the same year, a temperature of 1°.6 C. was registered in 1525 fathoms, immediately south of the Josephine Bank.

If we draw a line connecting the western extremities of the plateaux of the Cape de Verde Islands, the Canary Islands, and Madeira, it will be found that the isotherms of the stations west of this line are lower than the isotherms of the stations east of this line, and that, consequently, the water between this line and the coast of Africa is colder than it is to the westward of the above islands. The presence of this cold under-current along the West Coast of Africa, moreover, would be a necessary consequence of the general law which governs the circulation of thermal currents. The quantity of cold water supplied to the Atlantic by the Arctic basin is much less than the quantity flowing into it from the Antarctic region, and the principal supply is undoubtedly derived from the latter source. The cold bottom-current traced by H.M.S. "Challenger" from the Falkland Islands along the east coast of South America as far as Cape S. Roque, and thence across to St. Paul Rocks, and along the southern slope of the equatorial plateau as far as Station 346, with the bottom-temperatures slowly rising from $-0°.4$ C. to $+0°.4$ C., in this long course of several thousand miles, proves that there is such an inflow of cold water from the south towards the north. Such a current, after crossing the equator, would have a tendency to flow eastward and press up against the coast of Africa, owing to the same cause which compels the Arctic current to press up against and to flow along the coast of North America. Unfortunately, we do not possess on the African coast such a complete series of temperature-soundings as that furnished by the officers of the U.S. Coast Survey, and future observations can alone show how far the above theoretical conclusions are supported by facts.

The portion of the Atlantic section from Stations 84 to 95 is a continuation southwards of the eastern basin represented in the section from Teneriffe to Sombrero, Station 1 to Station 10. It is probable that the combined warm and cold currents flowing southwards from the Azores encounter, when off the Cape de Verde Islands, the cold equatorial stratum, and turn westwards, along the northern limit of the equatorial belt.

Before proceeding further, it may be necessary to allude to an appearance in the Atlantic section (Plate 9) which is also strongly marked in the Pacific section (Plate 19). It is the increasing undulations of the isotherms, as we enter the lower strata, at first sight suggesting the completely erroneous idea that the fluctuations of temperature increase with the depth, the very opposite being the case. This appearance is due partly to the excess of the vertical over the horizontal scale, the former being to the latter as 1200 to 1 in the Atlantic section, and as 1500 to 1 in the Pacific section—a disproportion inevitable on account of the relative thinness of the aqueous envelope when compared to the diameter of the areas which it occupies. It arises partly also from the nature of a section composed of *isotherms*. A difference of 0°.1 C. in temperature, which in the higher strata corresponds to only a few fathoms, and causes no perceptible alteration in the form of the isothermal line, is represented by 50 or 100 fathoms in the lower strata, and causes the isothermal line to assume the shape of a big wave. This defect, which requires only to be known to be guarded against, is obviated by the employment of *isobathymetrical* lines. As the latter represent the fluctuations of temperature at the same depth along the section of an oceanic basin, their undulations will keep pace with the fluctuations of temperature, greater near the surface, smaller as we enter the lower strata. But as in such a diagram the scale of depth must be replaced by a scale of temperature, it ceases to be a representation of an actual section (all but the horizontal distances) of an oceanic basin, and therefore the employment of a diagram of isotherms is, for general purposes, to be preferred. Of course, where space will allow it, both kinds of thermometric sections may be used, as has been done in the published diagrams of the U.S. Coast Survey.

SECTION FROM CAPE PALMAS TO CAPE S. ROQUE (Plate 10, Table V.).—The principal feature of this equatorial section has already been discussed above. A comparison of the isotherms between Station 106 and Station 102, and of the isotherms between Stations 339 and 342 of the Atlantic section (Plate 9), with the isotherms between Stations 129 and 119 of the southern section (Plate 10), shows that the north-eastern portion of the South Atlantic is warmer than its north-western portion.

A line drawn from Station 106 to Station 339 and continued down to the latitude of Tristan d'Acunha, which is a line proceeding in a south-south-easterly direction from a point to the eastward of St. Paul Rocks and intersecting the meridian of Ascension Island on the parallel of St. Helena, will divide the South Atlantic into two halves—a colder western half as regards the temperature of the lower strata, and a warmer eastern half. The contrast is the same as that found in the North Atlantic, and arises from a similar cause—a warming of the lower strata in the eastern half caused by the return eastward as an undercurrent of the South Atlantic Equatorial Current. There exists probably an additional cause of the higher temperature of the lower strata of the eastern half of the South Atlantic in comparison with the western half, namely, a return under-current of warmer water from the North Atlantic, which, finding the depths of the western half occupied by the Antarctic Current, flows down the eastern valley of the South Atlantic between the central plateau and the coast of South Africa.

We possess, in this part of the world, only a few soundings of the German frigate "Gazelle," which, however, prove the higher temperature of the lower strata in the eastern half of the South Atlantic, but further observations between Ascension, St. Helena, and the Cape of Good Hope are necessary.

SECTION BETWEEN CAPE S. ROQUE AND TRISTAN D'ACUNHA (Plate 10, Table V.).—This section passes in a south-easterly direction from Cape S. Roque to Tristan d'Acunha. The arrangement of its isotherms marks the progress of the South Atlantic Equatorial Current along the coast of South America, and especially of that branch of it which, turning eastwards, occupies the space between Station 339 and Tristan d'Acunha (Plate 9), and between Stations 131 and 132 (Plate 10). As seen in both sections, this branch attains its maximum depth between lat. $30°$ and $35°$ S. A comparison of the isotherm of $2°.5$ C. in the two sections shows that, while in the western half of the South Atlantic it is found not lower than at 1000 fathoms from the

TABLE V.—TEMPERATURES OBSERVED BETWEEN C. PALMAS, C. S. ROQUE, & TRISTAN D'ACUNHA—*August, October, 1873.*

Station No.	135	134	132	131	130	129	119	116	112	110	106	105	104	103	102
Latitude and Longitude	37° 20' S. 12° 20' W.	36° 12' S. 12° 16' W.	35° 25' S. 23° 40' W.	29° 9' S. 28° 6' W.	26° 15' S. 32° 56' W.	20° 13' S. 35° 19' W.	7° 39' S. 34° 12' W.	5° 1' S. 33° 50' W.	3° 33' S. 32° 16' W.	0° 0' N. 30° 18' W.	1° 47' N. 24° 26' W.	2° 6' N. 22° 53' W.	2° 25' N. 20° 1' W.	2° 49' N. 17° 13' W.	3° 8' N. 14° 49' W.
Surface Temp. C.	12°.0	12°.8	14°.4	18°.3	20°.6	23°.3	25°.3	25°.6	25°.6	25°.3	26°.0	25°.6	25°.6	25°.0	25°.6
Isotherm of 25° / 77° F.	—	—	—	—	—	—	—	—	30	30	35	—	30	0	15
Isotherm of 20° / 68° F.	—	—	—	—	—	60	60	40	45	50	45	—	40	45	45
Isotherm of 15° / 59° F.	—	—	—	130	100	130	125	60	60	75	60	80	75	90	100
Isotherm of 10° / 50° F.	65	130	200	270	260	220	180	150	160	165	150	145	190	160	210
Isotherm of 5° / 41° F.	360	350	435	420	400	355	300	260	320	325	260	400	460	360	360
Isotherm of 2°.5 / 36°.5 F.	600	800	850	1000	1000	950	1400	1200	1500	1300	1500	—	1250	—	1250
Bottom Temp.	—	1°.6	1°.1	0°.7	0°.8	0°.6	2°.3	0°.7	0°.5	0°.9	1°.8	1°.8	1°.7	1°.6	1°.7
Depth in Fms.	1000	2025	2050	2275	2350	2150	1650	2275	2200	2275	1850	2275	2500	2475	2450

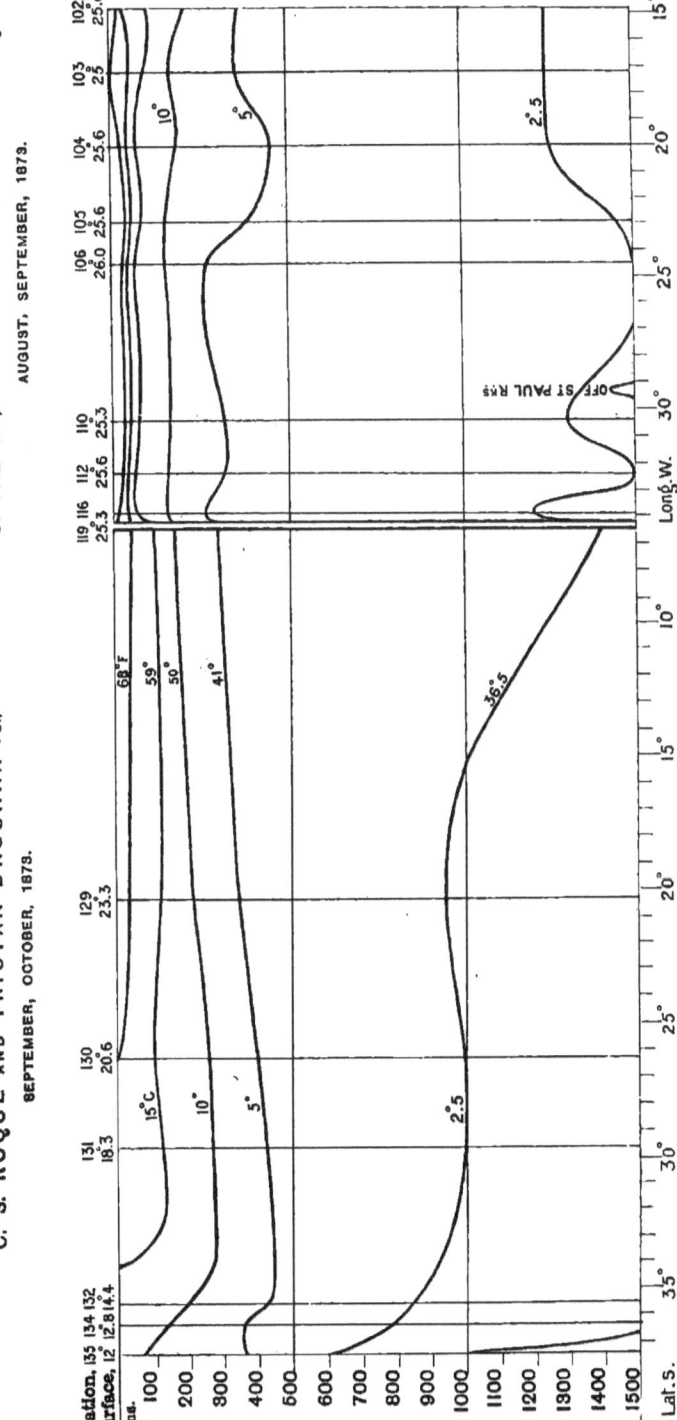

Plate 10.

surface as far north as lat. 15° S., it sinks below that level in lat. 30° S. between Stations 335 and 336 (Table IV.), upon the plateau which divides the western from the eastern half of the South Atlantic—a further proof of the higher temperatures which prevail in the lower strata of the latter as compared with the former. At Station 340, towards Ascension, the isotherm of 2°.5 C. is already below 1500 fathoms, and its remaining near that level as far as the equator indicates the presence of a large accumulation of warm water in the depths of the eastern half of the South Atlantic between lat. 20° S. and the equator. This circumstance has suggested the existence of a submarine ridge connecting the Central Atlantic plateau with the coast of Africa between the parallels of lat. 20° and 35° S., as shown in the chart of Staff-Commander T. H. Tizard which accompanies No. 7 of the *Report on Ocean Soundings*, by Captain Frank T. Thomson of H.M.S. "Challenger," and published by the Hydrographic Office. There are indications of the existence of such a ridge or area of elevation furnished by the discovery of several shallow soundings of less than 2000 fathoms between the central plateau and the coast of Africa, but the reasons stated above perhaps suffice to explain the presence of higher temperatures in the eastern basin of the South Atlantic, the more so as we observe a similar phenomenon in the North Atlantic. We know, besides, that a current of cold water opposes as effectual an obstacle to the further extension of warmer strata as a solid barrier formed by a submarine ridge or protuberance of the earth's crust. Further soundings in this region of the South Atlantic will decide this question.

SECTION FROM THE FALKLAND ISLANDS TO THE CAPE OF GOOD HOPE (Plate 11, Table VI.).—This section includes the Stations 317, 318, 319, and 320, situated between the Falkland Islands and the mouth of the Rio de la Plata, and they were added as belonging to the same thermal area. The section between

TABLE VI.—TEMPERATURES OBSERVED BETWEEN THE FALKLAND ISLANDS, TRISTAN D'ACUNHA, AND CAPE OF GOOD HOPE—*February, March, 1876; October, 1873.*

Station No.	317	318	319	320	323	324	325	326	327	329	330	331	332	333	334	135	136	137	138	139	140
Latitude and Longitude.	48° 37′ S. 55° 17′ W.	42° 32′ S. 56° 27′ W.	41° 54′ S. 54° 48′ W.	37° 17′ S. 53° 52′ W.	35° 39′ S. 50° 47′ W.	36° 9′ S. 48° 22′ W.	36° 44′ S. 46° 19′ W.	37° 3′ S. 44° 17′ W.	36° 48′ S. 42° 45′ W.	37° 31′ S. 36° 7′ W.	37° 45′ S. 33° 0′ W.	37° 47′ S. 30° 20′ W.	37° 29′ S. 27° 31′ W.	35° 36′ S. 21° 12′ W.	35° 45′ S. 18° 31′ W.	37° 20′ S. 12° 20′ W.	36° 43′ S. 7° 13′ W.	35° 59′ S. 1° 34′ E.	36° 22′ S. 8° 12′ E.	35° 8′ S. 16° 8′ E.	35° 0′ S. 17° 57′ E.
Surface Temp. C. F.	8°.2	14°.2	15°.3	19°.7	23°.0	22°.0	21°.6	19°.9	21°.2	18°.0	17°.9	18°.0	17°.8	19°.4	20°.3	12°.0	12°.2	13°.4	13°.4	13°.4	14°.6
Isotherm of 25° 77°	—	—	—	—	—	—	—	—	—	—	—	—	—	—	—	—	—	—	—	—	—
20° 68°	—	—	—	—	—	—	—	—	25	—	—	—	—	—	0	—	—	—	—	—	—
15° 59°	—	—	5	30	50	55	35	0	150	35	50	40	25	40	50	—	—	—	—	—	—
10° 50°	30	40	40	45	165	110	140	40	265	160	225	210	220	165	200	65	180	220	200	105	—
5° 41°	500	65	125	130	300	200	255	125	370	260	380	345	340	320	365	360	340	370	370	300	—
2°.5 36°.5	—	100	800	650	390	300	370	220	900	900	900	900	1000	850	900	600	700	800	—	800	—
Bottom Temp.	1°.7	0°.3	-0°.4	2°.7	0°.0	-0°.4	-0°.4	-0°.4	-0°.3	-0°.6	-0°.3	1°.3	0°.4	1°.2	1°.5	—	1°.1	0°.7	1°.0	0°.5	—
Depth in Fms.	1035	2040	2425	600	1900	2800	2650	2775	2900	2675	2440	1715	2200	2025	1915	1000	2100	2550	2650	2335	1250

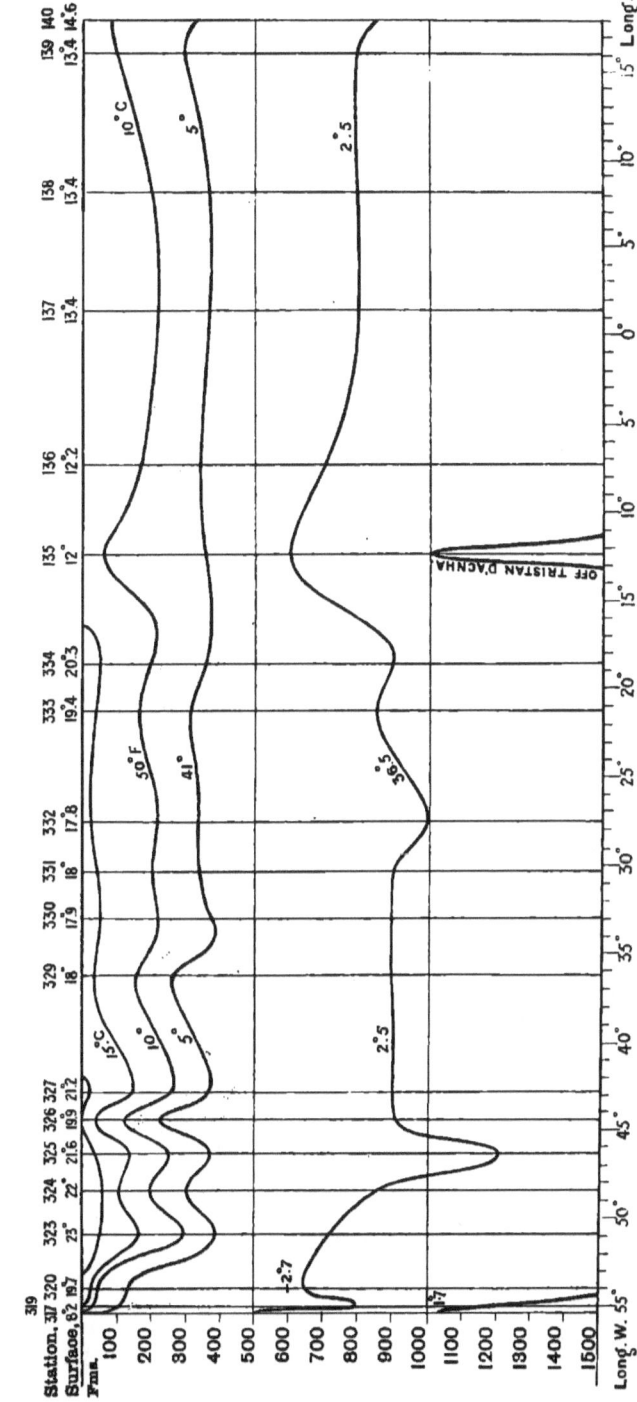

Station 323 and Station 140 traverses the whole of the South Atlantic Ocean from the mouth of the Rio de la Plata to the Cape of Good Hope, between the parallels of lat. 35° and 38° S., and only a few degrees north of what may be considered as the limit between the South Atlantic and the Southern Ocean. It is a combination of two sections surveyed at two different periods of the circumnavigation cruise of H.M.S. "Challenger," and Station 333 of the homeward voyage in March, 1876, nearly coincides with Station 133 of the outward voyage in October of 1873. The former date being in those southern latitudes the end of summer, when warm currents may be expected to have attained their maximum volume and velocity, while the latter date marks the beginning of spring, when cold currents have acquired their greatest power, this difference in the seasons ought not be entirely lost sight of in a comparison of the distribution of temperature in the two portions of the section east and west of Tristan d'Acunha. At Stations 317, 318, 319, and 320, we find the Antarctic Current pressing up against the coast of Patagonia and nearly coming to the surface at Station 317; while at Stations 318, 319, and 320, it is disguised by a warm surface-stratum about 100 fathoms thick—an extension or overflowing of the South Atlantic Equatorial Current between Rio de la Plata and the Falkland Islands. An examination of Curve Fig. 7, belonging to Station 318, shows that the Antarctic current here forms a stratum of the enormous thickness of about 1400 fathoms, or one and a-half English miles, with a nearly uniform temperature of from 1° to 2° C. The abnormal bottom-temperatures ascertained by the "Challenger" (varying from $-0°.3$ to $-0°.6$ C.), between Station 319 and Station 330 (long. 55° W., and long. 33° W.), a distance of over 1000 nautical miles, have been alluded to in a former chapter, as well as the slight increase of these temperatures towards the equator, where they are still found to vary between $0°.4$ C. and $0°.9$ C.—an increase of less than 1° C. in a distance of about 40° of latitude, or 2400 nautical miles.

Between Station 320 and Station 326, a distance of about 460 miles, we traverse the South Atlantic Equatorial Current, forming a surface-stratum of an average depth of 50 fathoms, with a temperature above 20° C. The undulating form of the isotherms marks the struggle going on between the equatorial and the polar current, betraying itself at the surface, as in the case of the Gulf Stream or North Atlantic Equatorial Current, by the formation of alternate streaks of warm and cold water. The cold under-current which presses up at Station 324 and Station 329 comes to the surface at Station 326. The warm current predominates at Stations 323, 325, 327 (at which latter station it forms a warm surface-streak beyond the cold streak of Station 326), and 330. As far as Station 334 we trace the warming influence of the equatorial current, a branch of which we have seen (Plate 9) flows to the northward of Tristan d'Acunha between the parallels of lat. 30° and 35° S. We have no observations in a direct line between Cape Horn and the Cape of Good Hope, but the main portion of the equatorial current seems, after flowing in a southerly direction from the place which it occupies in our section, to bend round between the parallels of lat. 40° and 50° S., and, coming in conflict with the Antarctic surface-current which flows towards the Cape of Good Hope, to sink under it and continue its course into the Antarctic regions as a warm under-current. The wide open sea, discovered by Weddell in 1823 beyond the parallel of lat. 70° S. and in the meridian of South Georgia (long. 40° W.), is probably an effect of this warm under-current.

The low surface-temperatures between Station 135 and Station 140 are due to the Antarctic surface-current which, flowing in a north-easterly direction, passes up in the space between Tristan d'Acunha and the Cape. The rise of the isotherms of 10° C. and of 2°.5 C. over the plateau of Tristan d'Acunha and Gough Islands deserves notice. The depth of the isotherms of 10° C. and 5° C. in the eastern portion of this section differs but little from that in the western portion. The concave shape of these isotherms between Tristan d'Acunha and the Cape is due probably to the influence of a comparatively warm surface-stratum, which, however, soon comes in contact with the cold bottom-stratum, for at Station 136, 4° C. was registered at 400 fathoms; at Station 137, 2°.6 C. at 700 fathoms; at Station 138, 3°.3 C. at 500 fathoms; and at Station 139, 3°.1 C. at 400 fathoms. The high level of the isotherm of 2°.5 C., rising in this section to 600 fathoms from the surface, while in the North Atlantic its position is generally at 1500 fathoms, proves the inflow of cold water from the Southern Ocean into the Atlantic basin, and the gradual rise of the temperature of the bottom strata as we proceed along the meridian from South to North.

CHAPTER V.

THE TEMPERATURE SECTIONS SURVEYED BY H.M.S. "CHALLENGER" IN THE SOUTHERN OCEAN, THE INDIAN ARCHIPELAGO, AND THE PACIFIC.

From the Cape of Good Hope to Melbourne—From Kerguelen Land to the Ice-barrier—From Sydney to Cook Strait, New Zealand—From Cook Strait to Tonga Tabu—From Tonga Tabu to Torres Strait—From Torres Strait to Hong-kong, and from Hong-kong to the Admiralty Islands—From the Admiralty Islands to Japan—From Yokohama to Station 253—From Station 253, in the Meridian of Honolulu and Tahiti, to Station 288—From Station 288 to Valparaiso and Magellan Straits.

SECTION FROM THE CAPE OF GOOD HOPE TO THE ICE-BARRIER AND TO MELBOURNE (Plates 12 and 13, Table VII.).—Much of the interest attached to the voyage round the world of H.M.S. "Challenger" centres in her cruise in the Southern Ocean. The latter, already associated with the fame of such great navigators as Cook (1773), Bellingshausen (1820), Weddell (1823), Morrell (1823), Biscoe (1831), Kemp (1834), Balleny (1839), D'Urville (1840), Wilkes (1840), Ross (1841 and 1843), and Moore (1845), had been the scene of many a courageous attempt to solve the mystery of the South Polar region, and it was the good fortune of Captain Sir George S. Nares, the leader of the recent Arctic expedition, to add his name to this long list of hardy explorers. It was not the mission of the "Challenger" to penetrate into the ice-bound regions of the South Pole—a task for which her size and her unprotected hull rendered her unfit—but she was able, thanks to the skilful navigation of her captain and officers, and with the assistance of a picked crew of England's sailors, to extend her dredging and sounding operations beyond the limits which had baffled former navigators, to cross the Antarctic Circle at a point not touched

From Cape of Good Hope to Melbourne. 89

by them, and perhaps to show the direction in which a future attempt to penetrate towards the South Pole might be successfully accomplished.

The section represented on Plate 12 traverses the Southern Ocean—not in a direction due east, but forms two sides of a triangle, the apex of which just touches the Antarctic Circle (Plate 1). The western side of the triangle is formed by the track of the "Challenger" from the Cape of Good Hope to Kerguelen Land and the Ice-barrier; the eastern side by the track of the ship between the latter and Melbourne.

The following table gives the surface-temperatures observed while crossing the "Agulhas Current"—the great equatorial current of the Indian Ocean.

TEMPERATURES OBSERVED BETWEEN THE CAPE OF GOOD HOPE AND MARION ISLAND, AND IN THE AGULHAS CURRENT.

Date, 1873.	Station.	Latitude at Noon.	Longitude at Noon.	Hour.	Surface Temperature.	Observations.
Dec. 17	141	34° 33' S.	18° 34' E.	4 a.m.	18°.0 C.	Warm streak.
,, ,,	... ...	,, ,,	,, ,,	8 ,,	16°.4	
,, ,,	... ...	,, ,,	,, ,,	9 ,,	15°.6	
,, ,,	... ...	,, ,,	,, ,,	10 ,,	13°.0	Cold streak.
,, ,,	... ...	,, ,,	,, ,,	11 ,,	13°.3	
,, ,,	... ...	,, ,,	,, ,,	Noon	18°.3	
,, ,,	... ...	,, ,,	,, ,,	3 p.m.	19°.2	Warm streak.
,, 18	142	35° 20'	18° 40'	Midnight	17°.5	Cold streak.
,, 19	143	36° 48'	19° 24'	1 a.m.	19°.4	
		,, ,,	,, ,,	2 ,,	22°.2	Warm streak.
		,, ,,	,, ,,	Noon	22°.8	Agulhas
,, 20	... ...	38° 6'	19° 53'	During the 24 hours	22°.2 to 19°.4	current. Cold streak.
,, 21	... ...	40° 37'	23° 12'	2 a.m.	22°.2	Warm streak.
,, ,,	... ...	,, ,,	,, ,,	2.30 ,,	20°.6	
,, ,,	... ...	,, ,,	,, ,,	3 ,,	18°.3	
,, ,,	... ...	,, ,,	,, ,,	3.30 ,,	15°.8	
,, ,,	... ...	,, ,,	,, ,,	2 p.m.	15°.0	
,, ,,	... ...	,, ,,	,, ,,	Midnight	13°.9	Cold water.
,, 22	... ...	42° 21'	27° 58'	During the 24 hours	14°.2 max. 8°.9 min.	,, ,, ,, ,,
,, 23	... ...	44° 41'	31° 28'	,,	9°.0 max. 6°.1 min.	,, ,, ,, ,,
,, 24	144	45° 57'	34° 39'	,,	8°.9 max. 5°.3 min.	,, ,,
,, 25	... ...	46° 28'	36° 43'	,,	5°.7 max. 4°.2 min.	Off Marion Isld. Pr. Edward Isls.
,, ,,	... ...	,, ,,	,, ,,			

G

TABLE VII.—TEMPERATURES OBSERVED BETWEEN CAPE OF GOOD HOPE, THE ANTARCTIC CIRCLE, AND C. OTWAY, *December, 1873; March, 1874.*

Station No.	140	141	143	144	146	147	150	153	154	Feb.21	156	157	158	159	160
LATITUDE AND LONGITUDE.	35° 0′ S. 17° 57′ E.	34° 41′ S. 18° 36′ E.	36° 48′ S. 19° 24′ E.	45° 57′ S. 34° 39′ E.	46° 46′ S. 45° 31′ E.	46° 16′ S. 48° 27′ E.	52° 4′ S. 71° 22′ E.	65° 42′ S. 79° 49′ E.	64° 37′ S. 85° 49′ E.	63° 30′ S. 89° 6′ E.	62° 26′ S. 95° 44′ E.	53° 55′ S. 108° 35′ E.	50° 1′ S. 123° 4′ E.	47° 25′ S. 130° 32′ E.	42° 42′ S. 134° 10′ E.
Surface Temp. C.	14.6	19.2	22.8	6.1	6.1	5.0	3.1	−1.2	0.0	0.0	0.6	2.9	7.2	10.8	12.8
ISOTHERM OF 15° / 59° F.	—	—	50	—	—	—	—	—	—	—	—	—	—	—	—
10° / 50°	—	40	140	—	—	—	—	—	—	—	—	—	—	50	70
5° / 41°	—	85	380	120	50	0	—	—	—	—	—	—	200	600	500
2°.5 / 36°.5	—	360	—	900	850	400	50	300	100	—	—	50	550	900	750
0° / 32°	—	—	—	—	—	—	—	125	50	40	100	—	—	—	—
−1.5 / 29°.3	—	—	—	—	—	—	—	—	—	—	—	—	—	—	—
Bottom Temp.	—	—	1.4	1.7	1.5	0.8	1.8	—	0.5	—	—	—	0.3	0.8	0.2
Depth in Fms.	1250	—	1900	1570	1375	1600	150	1675	1800	—	1975	1950	1800	2150	2600

TEMPERATURES IN THE SOUTHERN OCEAN.
BETWEEN
C. GOOD HOPE, P. EDWARD Is., CROZET Is., KERGUELEN Is., HEARD Is., ANTARCTIC CIRCLE, AND C. OTWAY,
DECEMBER, 1873, TO MARCH, 1874.

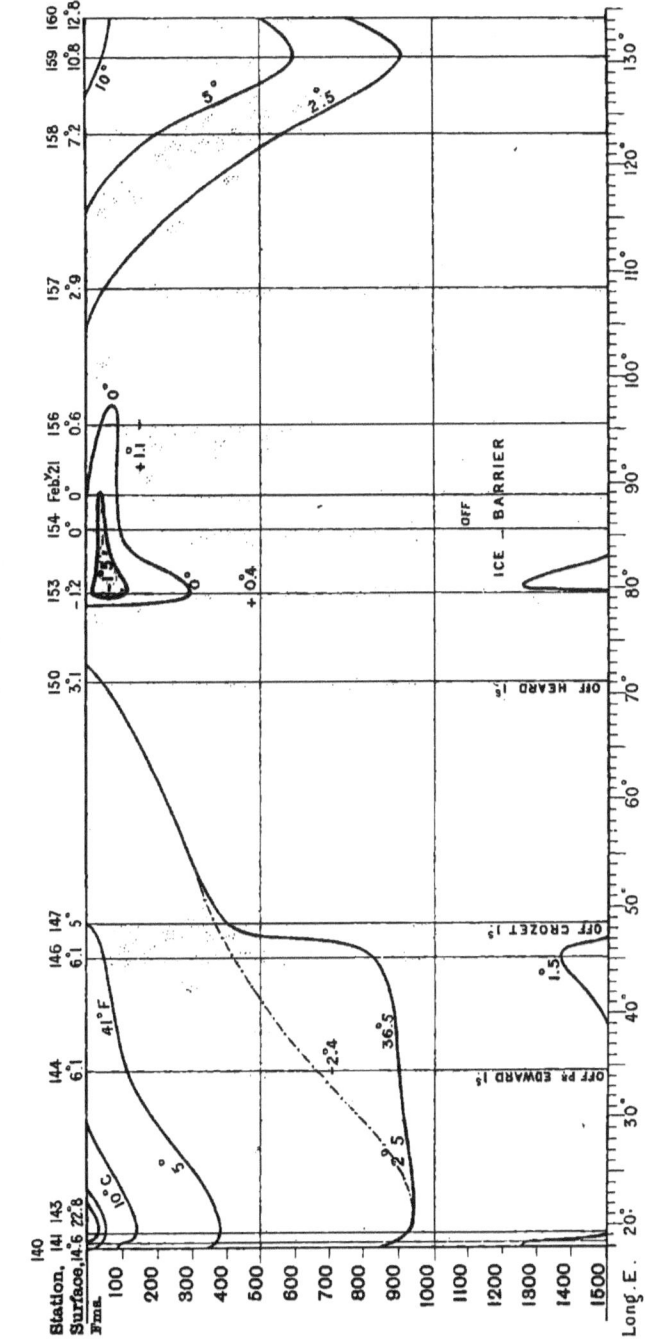

H.M.S. "Challenger" left her anchorage in Simons Bay at 6.30 a.m., the 17th December, 1873. Between 10 and 11 a.m., she traversed a current of cold water with a surface-temperature of 13° C., or 5° below the temperature recorded at 4 a.m. At noon the thermometer had again risen to 18°.3 C., and at 3 p.m. to 19°.2 C. After sailing at midnight on the following day through a second streak of cold water, the ship crossed the northern limit of the Agulhas Current between 1 and 2 a.m. of the 19th December, at which time the surface-temperature was observed to rise to 22°.2 C. Towards noon of the same day, and at Station 143, distant about 150 nautical miles from the Cape, the thermometer recorded the maximum of 22°.8 C. It next fell to 22°.2 C., and, with the exception of a cold streak of 19°.4 C. observed in the course of the 20th December, remained stationary until 2 a.m. on the 21st, between which hour and 2.30 a.m. it fell from 22°.2 C. to 20°.6 C. This, then, was the southern limit of the Agulhas Current, and the distance run between the two limits amounted to about 250 miles.

The existence of a streak of cold water in the opening of Simons Bay and in the immediate vicinity of the Cape, agrees with an explanation attempted by the author of the sudden changes of temperature observed in Simons Bay, and of the difference in the temperature of the water in the latter bay as compared with Table Bay. It is contained in a short paper published in *The Cape Monthly Magazine* for January, 1874, edited by the late Professor Noble, and its substance may be repeated here, as it furnishes a good illustration of the incessant contest going on between the equatorial current of the Indian Ocean and the polar current of the Southern Ocean, in the seas off the Cape of Good Hope.

The observations arranged in the following table were made on board H.M.S. "Challenger," supplemented by Surgeon

Thomas Bolster of H.M.S. "Flora," and kindly placed at the author's disposal by Captain Sir G. S. Nares.

Date, 1873.	Temperatures Observed in			
	Simons Bay.		Table Bay.	
	Temperature at 9 a.m.	Direction of the Wind.	Temperature at 9 a.m.	Direction of the Wind.
Nov. 27	11°.0 C.	N.W. by N.	—	—
„ 28	10°.6 C.	S.W.	—	—
„ 29	14°.0 C.	S.S.E.	—	—
„ 30	16°.4 C.	S.E. by S.	—	—
Dec. 1	16°.5 C.	S.E.	—	—
„ 2	15°.6 C.	N.W.	15°.6 C.	N.E.
„ 3	11°.1 C.	N.W.	12°.3 C.	N.N.W.
„ 4	12°.0 C.	S.W.	13°.0 C.	W.N.W.
„ 5	16°.7 C.	S.	12°.7 C.	S.
„ 6	16°.1 C.	S.E.	13°.9 C.	N.W.
„ 7	16°.7 C.	S.E.	12°.0 C.	S.S.E.
„ 8	16°.1 C.	S.E.	10°.8 C.	S.
„ 9	—	—	10°.3 C.	N.W.
„ 10	17°.8 C.	S.E.	10°.7 C.	N.N.E.
„ 11	17°.8 C.	S.E.	—	N.N.W.
„ 12	17°.5 C.	S.E.	—	—

The great surface-current of the Southern Ocean, as it flows from west to east, makes the circuit of the world between the parallels of lat. 60° and 40° S., and is split up by the projecting continents of South America, South Africa, and Australia into several branches, which can be traced flowing northwards along the western coasts of these continents, and which constitute the cold surface-currents of the South Pacific, the South Atlantic, and the Indian Ocean. The rapid fall of temperature—from 16° C. to 5° C. in the summer, from 10° C. to 5° C. in the winter of the southern hemisphere between the parallels of lat. 40° and lat. 50° S.—shows that the great surface-current of the Southern Ocean forms between these latitudes

From Cape of Good Hope to Melbourne.

an effectual barrier, a "cold wall," which arrests the further progress southwards of the equatorial currents of the three great oceanic basins. These warm currents—the Brazilian Current in the South Atlantic, the Agulhas Current or Cape Current of the Indian Ocean, and the currents which flow along the east coast of Australia and of New Zealand—on meeting the easterly current of the Southern Ocean, are split up into two, or rather three portions: the first portion is bent round and flows eastwards as a warm surface-current; the second, mixing with the cold current, is also carried eastwards, and accounts for the rise of temperature—from $5°$ C. to $10°$ C. or $16°$ C. according to the season—which is observed between lat. $50°$ and $40°$ S.; the third sinks below the cold surface-current, and, taking a south-easterly course, flows as a warm under-current into the Antarctic Ocean. These are the warm currents which undermine the enormous ice-masses that rise, under the name of the "Ice-barrier," like a solid wall to a height of from 150 to 300 feet above the surface of the sea, and detach from them the innumerable floating icebergs which strew the face of the Southern Ocean down to $50°$ and $40°$ latitude. The continual melting of these icebergs between $60°$ and $40°$ latitude supplies the masses of cold water which, to a depth of several thousand fathoms, fill up the basins of the Atlantic, the Pacific, and the Indian Ocean.

This splitting-up of the currents assumes especially marked features off the Cape of Good Hope. The Agulhas Current, immediately after crossing the meridian of the Cape, flows into the angle between the two branches of the Antarctic Current, between which it is completely annihilated *as a surface-current*, for few if any traces of it appear at the surface further westward. At the time of the "Challenger's" visit to these latitudes, the surface-temperatures between Tristan d'Acunha and the Cape ranged from $12°$ C. to $15°$ C., while the temperature of the surface-stratum of the Agulhas Current to a depth of 20 fathoms was found to be 22°.2 C. A portion of this current, however, may be found in the shape of an under-current for a distance of about 150 miles to westward of the Cape, as will appear on comparing the two Curves A and B, Fig. 5. The undulating form of Curve B indicates, as pointed out on a previous occasion, the presence of contending currents, which, being split asunder during the encounter, form alternate currents of warm and cold water flowing side by side and one above the other.

A portion of the Agulhas Current is probably carried by the Antarctic Current past the Cape and northward along the west coast of the African continent, but by far the greater portion is turned back and flows eastward, partly mixing with the cold current of the Southern Ocean, partly as an under-current. The latter seems to divide off the Crozet Islands into two branches, one of which continues its eastward course, while the other flows in a south-easterly direction towards the opening between Kemp Land and Wilkes' Termination Land, explored by H.M.S. "Challenger."

Near the Cape of Good Hope the forces of the Agulhas Current and of the Antarctic Current are so evenly balanced that, as appears from the above table, False Bay, which includes Simons Bay, is alternately occupied by branches of the warm or of the cold current, according as the wind blows from the south-east or from the north-west. During the prevalence of the former wind, a warm current is observed to flow from Cape Agulhas towards Cape Point, or from east to west, and False Bay is occupied by a branch of the Agulhas Current; during the prevalence of the latter, a cold current circles round the bay from west to east, and the bay is taken possession of by the Antarctic Current. A prevailing north or north-west wind drives the warm water out of False Bay, the bottom of which, from a line drawn between the Cape of Good Hope and Cape Hangklip, gradually shelves up from a depth of 50 fathoms. A prevailing south or south-east wind brings the branch of the Agulhas Current which flows over the Agulhas Bank into False Bay, raising the temperature of the water in Simons Bay 6° or 7° C. (11° to 13° F.). This difference was observed not only at the surface but at the depth of 9 fathoms, in which the "Challenger" was anchored, and the change would be accomplished in the short space of six hours.

The peninsula, from Table Mountain to the Vasco de Gama Hill, must at one time have been an island about thirty miles long and five miles broad, now joined to the mainland of Africa through the slow silting up of the strait which formerly flowed between Table Bay and False Bay.

Eastward of the Crozet Islands, the temperature of the Southern Ocean decreases rapidly. The isotherm of 5° C., which at Station 146 is at 50 fathoms, rises to the surface at Station 147. The temperature at 100 fathoms, which at the latter station is 2°.9 C., falls to 1°.8 C. at Station 150, and to 0°.0 C. at Station 152. On the morning of the 11th February, and in about lat. 60° 40′ S., long. 80° 20′ E., the "Challenger" sighted the first iceberg. At 4 a.m. it presented the appearance of a silvery mass dimly visible towards the east-south-east, and was found, by angular measurement, to be over 700 yards long, and to rise vertically on all sides to a height of more than 200 feet above the surface of the water. From that date to the end of the month, the narrow horizon commanded from the ship's deck offered the imposing spectacle of a sea studded with icebergs of every size and shape, though generally assuming the form of huge slabs, whose snow-white surface reflected every hue of day—from the delicate silvery-grey of dawn to the golden and crimson tints of sunset, and from the inky darkness of their recesses when in the shadow of a cloud, to the exquisite azure-blue light which filled the numerous caves worn by the waves in their flanks. On the day when the ship stopped in lat. 66° 40′ S., long. 78° 22′ E.—a distance of 1400 nautical miles from the South Pole—more than seventy of these floating islands of ice could be counted from her deck, one of them over five miles long, and rising from 150 to 200 feet above the sea.

The melting of these ice-masses produces a quantity of water, which, being fresher, is of less specific gravity than the salt water of the surrounding sea, and therefore floats in the immediate vicinity of the ice on the surface of the latter. But as the fresher water derived from the icebergs mixes by degrees with the surrounding salt water, the mixture being of lower temperature is rendered heavier, and sinks below the surface, forming an intermediate stratum or wedge, as shown in Plates 12 and 13.

Owing to her supplies of coal running short, H.M.S. "Challenger" was prevented from establishing as many stations between the Ice-barrier and Australia as might have been desirable, but sufficient observations were secured to confirm the existence and to ascertain the proportions of the great current which, under the name of the South Australian Current, was already known to flow in an easterly direction at some distance from the south coast of the Australian continent.

The temperature of the water, which at Station 157 was 2°.9 C. at the surface, 2°.6 C. at 60 fathoms, 0°.6 C. at 70 fathoms, and 0°.3 C. at 80 fathoms—thus betraying the presence of an overflowing warmer current—and −0°.6 C. at the bottom in 1950 fathoms, had risen at Station 158 to 7°.2 C. at the surface, 5° C. at 200 fathoms, 2° C. at 700 fathoms, 1° C. at 1500 fathoms, and 0°.3 C. at the bottom in 1800 fathoms. At Station 159 there appeared evident signs of the presence of a warm under-current, for the temperature of the water, which between the surface and 100 fathoms had fallen from 10°.8 C. to

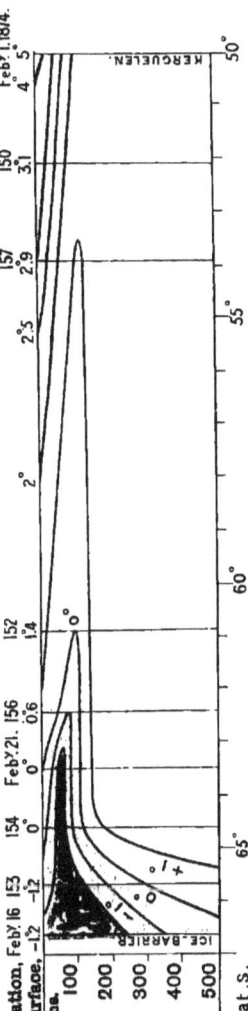

9°.3 C., remained almost stationary between 150 fathoms and 400 fathoms, descending from 8°.8 C. to 8°.2 C. at the latter depth, whence it decreased more rapidly to 5° C. at 600 fathoms, to 3° C. at 800 fathoms, to 2° C. at 1100 fathoms, and to 0°.8 C. at the bottom in 2150 fathoms. Similar conditions of temperature were observed at Station 160, nearer to the Australian coast, where, however, the warm stratum, still commencing with 8°.8 C. at 150 fathoms, ended with 8°.2 C. in 300 fathoms, while 5°.1 C. were registered at 500 fathoms, 2°.4 C. at 800 fathoms, and 0°.2 C. at the bottom in 2600 fathoms. It seems, therefore, that at the latter station the ship had already crossed the axis of this warm under-current.

The distance run between Station 158 and Station 159, and between the latter station and·Station 160, was about 350 miles, so that this current cannot be less than 400 miles broad. The distance between Station 160 and Cape Northumberland, the nearest point on the south coast of Australia, is about 380 nautical miles, and the axis of the current may be laid down on the chart at a distance of about 500 miles from the Australian coast, bending round in a south-easterly direction towards the wide space of open water at the foot of Victoria Land, discovered by Sir James Ross. It may be taken for granted that the South Australian current just described, as well as the currents to the westward of the Crozet Islands and Kerguelen Land, represent the outflow of the warm water of the Indian Ocean through the Southern Ocean into the Antarctic Basin.

SECTION FROM SYDNEY TO COOK STRAIT, NEW ZEALAND, AND FROM COOK STRAIT TO TONGA TABU (Plate 14, Table VIII.).— Narrow as the sea between Australia and New Zealand seems when compared with its great neighbour the Pacific, its average width is a thousand miles, and a ship sailing at the rate of 150 miles a-day will consume a week in accomplishing the voyage from Sydney to Wellington. Divided from the basin of the Pacific

TABLE VIII.—TEMPERATURES OBSERVED BETWEEN N. S. WALES, NEW ZEALAND, AND FRIENDLY ISLANDS—*June, July, 1874.*

Station No.	Latitude and Longitude	Surface Temp.	Isotherm of 25° C. / 77° F.	20° / 68°	15° / 59°	10° / 50°	5° / 41°	2°.5 / 36°.5	Bottom Temp.	Depth in Fms.
163	36° 56′ S. 150° 30′ E.	22°.2	—	30	115	235	—	—	0°.7	2200
164 A	34° 19′ S. 151° 31′ E.	20°.0	—	0	130	280	390	—	—	400
164 B	34° 27′ S. 154° 57′ E.	17°.8	—	—	60	160	480	850	—	2550
165 A	36° 41′ S. 158° 29′ E.	17°.0	—	—	60	220	520	900	0°.6	2600
165 B	37° 53′ S. 163° 18′ E.	15°.3	—	—	40	230	560	900	0°.9	1975
165 C	38° 36′ S. 166° 39′ E.	14°.6	—	—	—	260	520	850	2°.0	1100
166	38° 50′ S. 169° 20′ E.	14°.7	—	—	—	270	—	—	—	275
168	40° 28′ S. 177° 43′ E.	14°.0	—	—	—	245	650	1000	2°.0	1100
169	37° 34′ S. 179° 22′ E.	14°.6	—	—	—	285	600	—	4°.2	700
170	29° 55′ S. 178° 14′ W.	18°.3	—	—	100	260	560	—	—	520
171	28° 33′ S. 177° 50′ W.	19°.2	—	—	—	—	—	—	4°.0	600
171 A	25° 5′ S. 172° 56′ W.	22°.2	—	100	200	265	500	850	0°.5	2900

TEMPERATURES IN THE SOUTH PACIFIC,

BETWEEN

P. JACKSON, N. S. WALES, AND COOK STR., NEW ZEALAND.

JUNE, 1874.

TEMPERATURES IN THE SOUTH PACIFIC,

BETWEEN

COOK STR., KERMADEC IS., AND FRIENDLY IS.,

JULY, 1874.

Ocean by a submarine plateau which rises to within 1000 fathoms of the sea surface, and unites Australia, New Zealand, New Caledonia, and Papua into a single area of elevation, it may be considered as forming a bight of the Southern Ocean (Plate 2). The cross section of this area presents the not unfrequent contrast of deep soundings and a comparatively rapid fall of the sea-bottom along its western boundary, and of shallow soundings and a slowly rising bottom towards the east. The western half of the basin is occupied by an area of depression of more than 2500 fathoms, or about three miles in depth, extending from the south point of Tasmania along the east coast of Australia as far as Great Sandy Island, where the coast turns towards the north-east. The eastern half forms a broad plateau, which ultimately rises above the level of the sea under the name of New Zealand.

A branch of the South Pacific Equatorial Current, after passing to the southward of the Fiji Islands and New Caledonia, strikes the Australian coast near Great Sandy Island, and, bending round, flows as a surface-current, known under the name of the East Australian Current, close along the shores of New South Wales. With the exception of this current, the whole of the basin between Australia and New Zealand is occupied by a branch of the South Australian Current, which, crossing the basin in a north-easterly direction, carries a portion of the East Australian Current along with it, and divides into two branches, one returning southwards along the west coast of Middle Island, while the other flows round the North Cape and probably down the north-east coast of North Island.

A comparison of the isotherms at Stations 165 and 166 west of New Zealand, and at Stations 168 and 169 east of the latter, seems to show that at the time of the "Challenger's" visit, which was in the winter of the southern hemisphere, the whole plateau of New Zealand was swept by a current from the south-west, whose temperature was somewhat raised through carrying the waters of the East Australian current along with it. No trace of a branch of the South Pacific Equatorial Current flowing along the east coast of New Zealand appears at Stations 168 and 169 of the section between Cook Strait and Tonga Tabu; but as these stations are near the coast, and properly belong to the plateau of New Zealand, it is possible that such a branch may be found further to the eastward.

The above section exhibits the usual contraction of the isotherms as we approach the tropics. The soundings of the "Gazelle" and of the "Tuscarora" have proved that a channel of more than 2000 fathoms in depth passes up between New Zealand and the Kermadec Islands in a north-westerly direction towards New Caledonia. Split into two branches by the plateau which supports New Caledonia and Loyalty Islands, the eastern branch continues in the same direction between the latter islands and the New Hebrides, and finally communicates with the basin which stretches from the New Hebrides as far as Torres Strait.

Station 171A, with a depth of 2900 fathoms, belongs to the area of depression discovered by the "Gazelle," and which has been traced and explored by the German expedition from the Samoan Islands across the South Pacific as far as the Strait of Magellan.

SECTION FROM TONGA TABU TO TORRES STRAIT (Plate 15, Table IX.).—The combined soundings of H.M.S. "Challenger" and of the U.S.S. "Tuscarora" show that the Fiji Islands occupy the centre of a plateau which comprises the Samoan Islands in the north-east, Tonga Tabu or the Friendly Islands in the east, the Kermadec group in the south, and the New Hebrides in the west. This plateau may be considered as the terminal knot which unites two extensive areas of elevation. One of these stretches from the Samoan group, first in a north-north-easterly direction through the Ellice and

Gilbert islands to the Marshall group, and afterwards in a due westerly direction through the Caroline Islands as far as the Pelew Islands. The other connects the New Hebrides with the Santa Cruz Islands, the Solomon Islands, New Ireland, New Britain, the Admiralty Islands, and Papua. The two areas of elevation, or ridges as we may call them, enclose an area of depression which extends from Fiji to the Philippines, and which was partially explored by the "Challenger" in the months of February and March, 1875, and by the "Gazelle" in June and July, 1875. The total length of this basin amounts to 3000 miles, and as the continent of Papua constitutes about one-half of its southern boundary, it may with some show of reason be distinguished by the name of the "Sea of Papua." To the southward of the last-described ridge we find another area of depression, which stretches from the New Hebrides to Torres Strait, and forms a continuation of the 2000-fathom channel between the New Hebrides and New Caledonia. This is the basin represented in the Section Plate 15. Bounded on the south by the shallow coral-sea whose numerous reefs crowd the space between Australia and New Caledonia, it forms an almost land-locked basin, communicating with the depths of the South Pacific only through the above-mentioned 2000-fathom channel, which also divides the Fijian plateau from New Caledonia and New Zealand. The almost uniform level of the isotherms between Fiji and Torres Strait, a distance of about 2000 miles, is the most prominent feature of this section. It was this basin—named, at the time of its discovery by the "Challenger," the "Melanesian Sea" (partly on account of the dark complexion of the natives of the islands by which it is surrounded, partly to revive a term used by the earlier geographers and navigators)—which furnished the first example of the distribution of temperature in a sea separated from the depths of the ocean by a submarine ridge or area of elevation. The

TABLE IX.—TEMPERATURES OBSERVED BETWEEN THE FIJI ISLANDS & TORRES STRAIT—*August, 1874.*

STATION NO.		184	183	182	180	179	178	177	176	175	174
LATITUDE AND LONGITUDE.		12° 8′ S. 145° 10′ E.	12° 42′ S. 146° 46′ E.	13° 6′ S. 148° 37′ E.	14° 7′ S. 153° 43′ E.	15° 58′ S. 160° 48′ E.	16° 47′ S. 165° 20′ E.	16° 45′ S. 168° 5′ E.	18° 30′ S. 173° 52′ E.	19° 2′ S. 177° 10′ E.	19° 10′ S. 178° 10′ E.
Surface Temp.		25°.3	25°.6	25°.8	26°.7	26°.1	26°.1	26°.0	25°.3	25°.3	25°.6
ISOTHERM OF	C. 25° / F. 77°	45	40	25	40	55	40	—	10	10	20
	20° / 68°	120	110	95	120	120	120	—	110	130	130
	15° / 59°	175	170	150	180	170	190	—	190	200	200
	10° / 50°	235	240	220	250	260	270	—	250	270	270
	5° / 41°	385	400	385	430	420	430	—	430	435	440
	2°.5 / 36°.5	800	800	850	900	930	830	—	—	830	—
Bottom Temp.		1°.8	1°.7	1°.4	1°.4	1°.4	1°.3	—	2°.0	1°.8	3°.7
Depth in Fms.		1400	1700	2275	2450	2325	2650	125	1450	1350	610

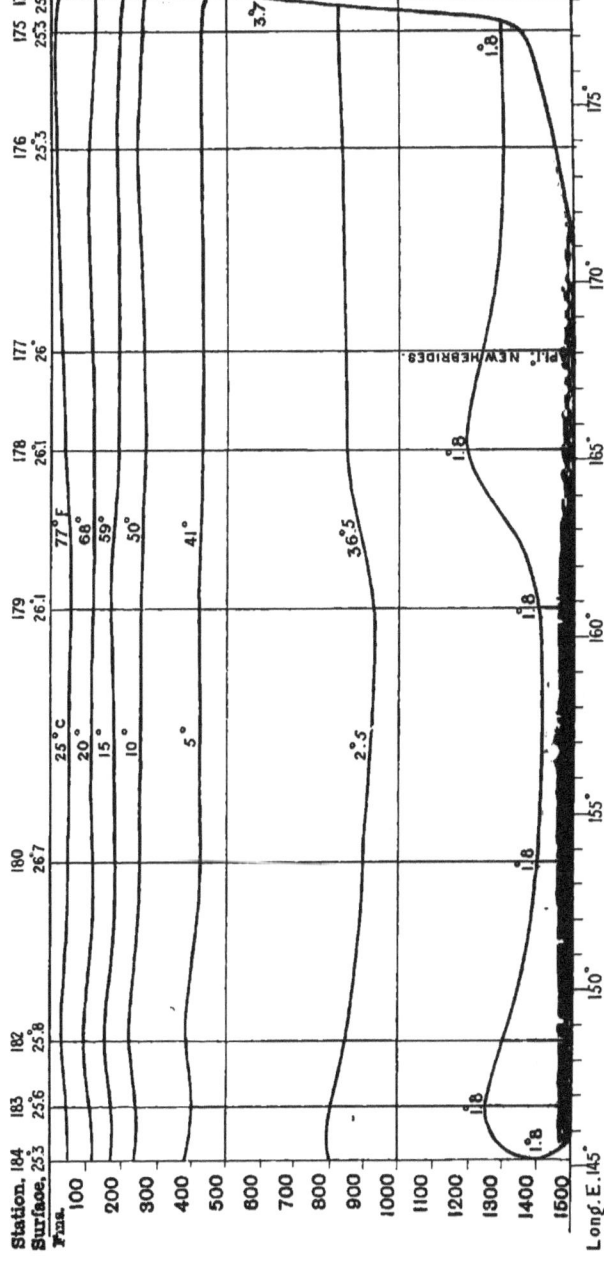

officers on board H.M.S. "Challenger" were surprised to find that, beyond a depth varying between 1200 and 1400 fathoms, the thermometers ceased to register any further decrease of temperature. The latter was observed to fall from 25° C. at the surface to 1°.8 C. between 1200 and 1400 fathoms, and to remain stationary or nearly stationary from that level down to the bottom. At Station 184, the bottom-temperature in 1400 fathoms was 1°.8 C.; at Station 183, in 1700 fathoms, 1°.7 C.; at Station 182, in 2275 fathoms, 1°.4 C.; at Station 176, between the Hebrides and Fiji, in 1450 fathoms, 2° C.; and at Station 175, in 1350 fathoms, 1°.8 C. The same phenomenon was found to occur under similar conditions in the seas of the Indian Archipelago, and tended to confirm the opinion formed at the time—that its cause must be sought in the partial or complete suspension of all deep-sea communication by intervening submarine ridges or submerged areas of elevation.

This circumstance affords one of the most convincing proofs of the existence of *a system of thermal circulation in a horizontal direction, which embraces the whole of the ocean as well as the minor seas in proportion to the facilities of submarine communication.* The almost uniform level of the isotherms observed in landlocked basins also shows *that the obliquity or gradient of the isotherms stands in direct relation to the presence or absence of currents of different temperature, and therefore of different origin, and moving in different directions in the various areas which compose the aqueous envelope of our planet.* In the same manner, it has been explained on a previous occasion *that the gradients of the temperature-curve stand in direct relation to the presence or absence of currents of different temperature, origin, and direction, in the various strata between the surface and the bottom.*

An examination of Table IX. shows that, although the

isotherms appear almost at the same level from one end of the section to the other in the diagram of Plate 15, the distribution of temperature varies from one station to another. There is a gradual decrease of temperature between 100 fathoms and 500 fathoms as we proceed westwards until we arrive at Station 182, which is the coldest, after which the temperatures rise again towards Station 184. The warm surface-stratum above 25° C. is only 10 fathoms deep between the Fiji Islands and the New Hebrides at Stations 175 and 176, a decrease probably due to the influence of a cold current from the southward, perhaps a branch of the cold current flowing northwards along the west coast of New Zealand. Its depth rapidly increases as we enter the Melanesian basin at the New Hebrides, attains a maximum of 55 fathoms at Station 179, falls to 25 fathoms at Station 182, and increases again to 40 and 45 fathoms as we approach Torres Strait. These alterations of level, which represent a difference in the thickness of the warm surface-stratum of over 100 feet, must be caused by warm and cold surface-currents flowing into the basin from the north or south.

SECTIONS FROM TORRES STRAIT TO HONG-KONG, AND FROM HONG-KONG TO THE ADMIRALTY ISLANDS (Plate 16, Table X.).—These sections embrace the numerous seas between Papua and China, and are chiefly interesting as offering several instances of the influence of submarine and surface barriers upon the distribution of temperature, the former interfering with the gradual decrease of temperature between the surface and the bottom, the latter resulting in a superheating of the surface-strata, which in these seas, as well as in others similarly circumstanced, attain a temperature not observed in the open ocean.

THE ARAFURA SEA.—Following the track of H.M.S. "Challenger," we enter the Arafura Sea through Torres Strait. The soundings taken between the latter and the Arrou Islands prove that the continents of Australia and Papua are bound together by a plateau about 600 miles broad and less than 50 fathoms below the sea-surface, which extends along the parallel of lat. 10° S. This plateau is separated from the Ki Islands, Timor Laut, and Timor by a channel varying from 1000 to 2000 fathoms in depth, which commences in the Indian Ocean, and passes close to the eastern shores of these islands. After separating Great Ki Island from the Arrou group, this channel bends round towards the west, flows between Ceram in the south, Mysole and Obi Major in the north, then, turning due north, it continues between Celebes and Gillolo under the name of the Molucca Passage, and enters the North Pacific near the Tulur Islands. The "Challenger" sounded in 1200 fathoms in the Molucca Passage, and in 800 and 580 fathoms on each side of the channel between Great Ki Island and Dobbo, Arrou Islands (Stations 191 and 191A). The "Gazelle" found 1720 and 995 fathoms north of Ceram. Older soundings mark depths of 1700 fathoms and 1070 fathoms off Timor.

This channel establishes what seems to be the only deep-sea communication between the Indian Ocean and the North Pacific, and has all the appearance of a "fault" on a gigantic scale (it is about 1500 miles long, and assumes the shape of S), separating the Papua-Australian plateau from the plateau of the Indian Archipelago. It is probably swept by powerful currents, and may have some connection with the remarkable contrast which the naturalist discovers between the fauna and flora of the regions east and west of it.

THE BANDA SEA.—The soundings of the "Gazelle" and "Challenger" of 2320 fathoms and 2800 fathoms show that this sea, notwithstanding its restricted area, covers a depression of considerable depth. The temperature of the water was found to decrease from 28°.6 C. at the surface to 10° C. at 200 fathoms, to 4° C. at 600 fathoms, and to 3° C. at 900 fathoms, from which depth it remained stationary down to the bottom in

H

TABLE X.—TEMPERATURES OBSERVED IN THE SEAS BETWEEN HONG-KONG AND NEW GUINEA OR PAPUA—*September, 1874, to March, 1875.*

Station No.	205	206	207	210	202	211	198	199	213	197	193	191	191A	214	215	216A	216B	217	218	220
Latitude and Longitude	119° 22' E. 16° 42' N.	117° 14' E. 17° 54' N.	122° 15' E. 12° 21' N.	123° 45' E. 6° 26' N.	121° 55' E. 8° 32' N.	121° 42' E. 8° 0' N.	124° 53' E. 2° 55' N.	123° 34' E. 5° 44' N.	124° 1' E. 5° 47' N.	126° 38' E. 0° 41' N.	130° 37' E. 5° 24' S.	134° 5' E. 5° 41' S.	133° 19' E. 5° 26' S.	127° 9' E. 4° 33' N.	130° 15' E. 4° 19' N.	133° 58' E. 2° 46' N.	134° 11' E. 2° 56' N.	138° 55' E. 0° 39' S.	144° 4' E. 2° 33' S.	147° 0' E. 0° 42' S.
Surface Temp. C. / F.	27°.8	24°.0	26°.7	26°.8	28°.6	27°.2	29°.4	28°.3	28°.3	28°.1	28°.6	27°.9	27°.5	27°.0	27°.7	28°.2	28°.2	28°.3	28°.9	28°.8
Isotherm 25° / 77°	25	—	25	55	40	35	50	65	55	35	55	30	45	35	15	75	85	90	85	100
20° / 68°	55	50	60	80	80	70	90	90	90	80	85	75	100	70	100	120	115	115	115	130
15° / 59°	110	100	95	110	130	110	120	120	120	125	130	140	150	105	160	160	145	140	150	175
10° / 50°	220	186	—	—	—	—	190	190	180	200	200	210	—	160	240	195	180	195	200	220
5° / 41°	530	460	—	—	—	—	435	—	500	475	470	510	—	550	500	470	—	460	420	600
2°.5 / 36°.5	900	880	—	—	—	—	—	—	—	940	—	—	—	—	875	875	—	866	1100	1150
Bottom Temp.	2°.4	2°.3	11°.0	12°.2	10°.2	10°.2	3°.7	3°.5	3°.0	1°.8	3°.1	3°.9	4°.9	5°.3	1°.1	1°.3	0°.9	1°.2	2°.1	2°.0
Depth in Fms.	1050	2100	700	375	2550	2225	2150	2600	2050	1200	2800	800	580	500	2500	1650	2000	2000	1270	1300

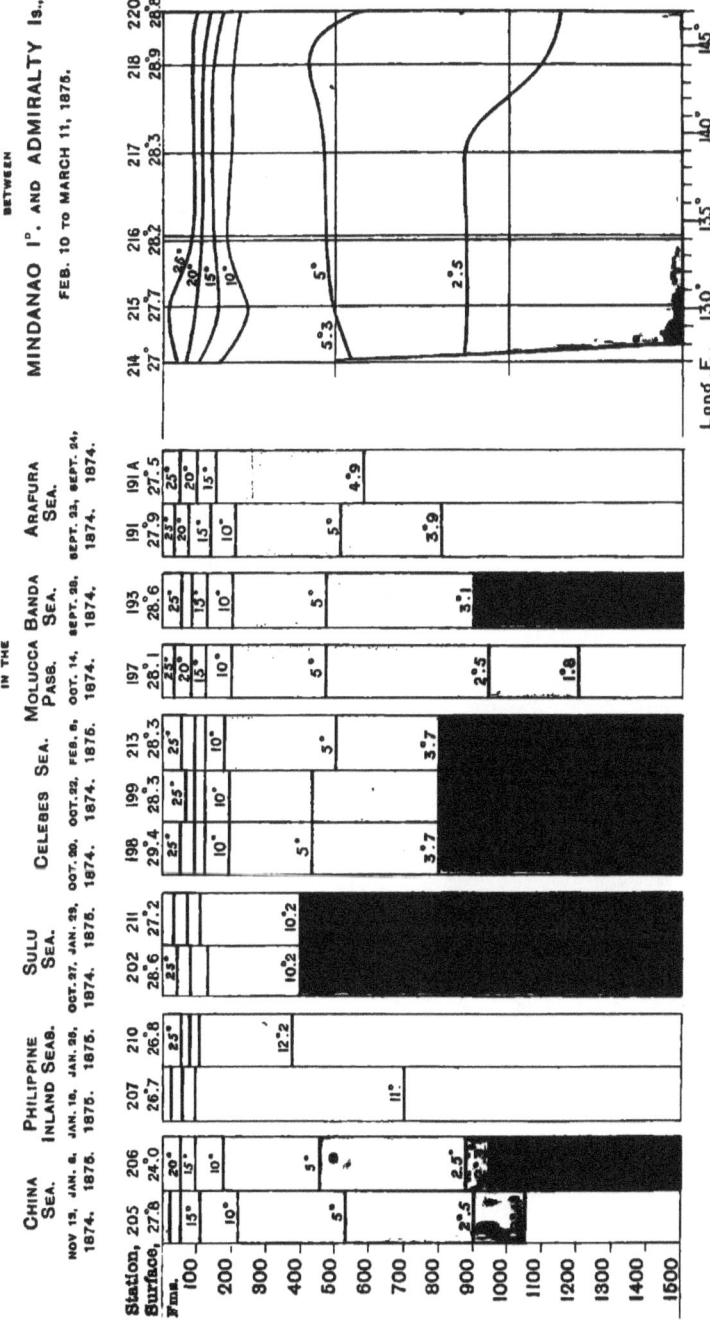

2800 fathoms, forming a stratum of more than two miles in thickness of the uniform temperature of 3° C. The existence of this stratum seems to prove that if the Banda Sea has any deep-sea communication with the Indian Ocean—of which there are certain indications in the deep soundings of 2005 and 2320 fathoms found by the "Gazelle" close to and north of the island of Timor—the depth of the channel or channels which connect the two oceans cannot exceed 900 fathoms. At Station 195, between Banda and Amboina, 4400 fathoms of dredge-rope were paid out in anticipation of a depth of 4000 fathoms, marked in the older charts, but the depth turned out to be only 1425 fathoms (bottom temperature, 3° C.)—a further proof of how little reliance can be placed on soundings taken with the imperfect appliances formerly in use.

THE MOLUCCA PASSAGE.—The bottom temperature of 1°.8 C., registered in a depth of 1200 fathoms at Station 197, contrasts remarkably with the temperature of the corresponding depth in the Banda Sea; and a comparison of the isotherms of this station with those of Station 214 and Station 215, at the entrance of the Molucca Passage, establishes the deep-sea communication between this passage and the North Pacific.

THE SEA OF CELEBES.—This sea forms another of those areas of narrow superficial extent, but of great depth, so characteristic of the Indian Archipelago. The several soundings of H.M.S. "Challenger" give depths of 2150, 2600, and 2050 fathoms, with bottom temperatures respectively of 3°.7 C., 3°.5 C., and 3° C. The decrease of temperature is arrested at a depth of 800 fathoms with a temperature of 3°.7 C. The line of islands between Celebes and Mindanao, therefore, forms an effectual barrier against the inflow of colder water from the Pacific.

THE SULU SEA.—The portion of this sea in the immediate vicinity of the east coast of Mindanao forms a deep hollow, with depths of 2225 and 2550 fathoms, and furnishes, with the Mediterranean, the most remarkable illustration of the conditions of temperature in a nearly land-locked basin. The two soundings of Station 202 and Station 211 were obtained at an interval of three months, the former in October, 1874, the latter in January, 1875. On both occasions the decrease of temperature ceased at a depth of 400 fathoms, and remained stationary at $10°.2$ C. from that level down to the bottom. It will be remembered that the temperature of the Mediterranean was found by H.M.S. "Porcupine" to remain stationary from 100 fathoms downwards.

THE PHILIPPINE INLAND SEAS.—The high bottom-temperatures observed in these seas prove that, like the Sulu Sea, they are virtually inland basins, and do not communicate with the China Sea in the west and with the Pacific in the east by channels deeper than 150 or 200 fathoms.

THE CHINA SEA.—This sea also forms a nearly land-locked basin. Its greatest depths as yet ascertained are situated in the north-eastern portion, near the island of Luzon. The observations made at Station 205 and Station 206 are about two months apart, which may account for the comparatively higher temperatures of Station 205. A typhoon had swept these seas in the interval. The decrease of temperature is arrested at a depth of about 1000 fathoms, below which the temperature of the water remains at $2°.4$ C. down to 2100 fathoms. This low temperature shows that the China Sea communicates with the Pacific by channels of considerable depth, probably situated between Formosa and Luzon.

THE SEA OF PAPUA.—Only the western portion of this large basin, which extends from the Philippines to the Fiji Islands, has as yet been explored—by the "Challenger" in February and March, 1875, and by the "Gazelle" in the course of June and July of the same year. The sections from Mindanao to the Admiralty Islands (Plate 16), and from the latter to Station 224 (Plate 17, Tables X. and XI.), represent the temperature-observations made

From the Admiralty Islands to Japan. 109

in this basin by the English expedition, the results of which show perfect agreement with those of the German expedition, making allowance for the difference of dates.

The position of Station 215, with a depth of 2500 fathoms, is exceptional. Placed in the centre between the Sea of Papua in the east, the Molucca Passage in the south, the Celebes Sea in the west, and the portion of the Pacific extending from the Philippines to the Pelew Islands in the north, it need cause no surprise that its isotherms should mark a considerable disturbance in the distribution of temperature, due to the presence of several currents of different temperature, direction, and origin. They give evidence of a colder current between the surface and 100 fathoms, and of a warmer current between 100 fathoms and 250 fathoms. The former may be traced to the north-west, and is probably a current from the east coast of Mindanao, the latter to the south-west and the Molucca Passage.

The remaining stations in this section belong to the thermal area of the Sea of Papua.

SECTION FROM THE ADMIRALTY ISLANDS TO JAPAN (Plate 17, Table XI.).—This section may be divided, both geographically and as regards distribution of temperature, into three parts. The first and most southern part, from Station 220 to Station 224 (including also Stations 216–220 of the previous section, Plate 16), forms the western extremity of the Sea of Papua, being bounded in the south by the latter continent, and in the north by the line of the Caroline and Pelew Islands. The second part, from Station 224 to Station 228, traverses the eastern limits of the sea situated between the Philippines and the Mariana or Ladrone Islands, and between the Pelew Islands and Japan. This sea, which fills up a deep basin separated from the rest of the Pacific by the line of islands extending from the Caroline Islands to Japan, has as yet received no name, and might appropriately be called the "Sea of Magallanes," after the discoverer of the Mariana

TABLE XI.—TEMPERATURES OBSERVED BETWEEN THE ADMIRALTY ISLANDS AND JAPAN—*March to June, 1875.*

STATION NO.	220	221	222	223	224	225	226	227	228	229	230	231	234	235	232
LATITUDE AND LONGITUDE	147° 0' E. 0° 42' S.	148° 41' E. 0° 40' N.	146° 16' E. 2° 15' N.	145° 13' E. 5° 31' N.	144° 20' E. 7° 45' N.	143° 16' E. 11° 24' N.	142° 13' E. 14° 44' N.	141° 21' E. 17° 29' N.	141° 13' E. 19° 24' N.	140° 27' E. 22° 1' N.	137° 57' E. 26° 29' N.	137° 8' E. 31° 8' N.	135° 39' E. 32° 31' N.	138° 0' E. 34° 7' N.	139° 28' E. 35° 11' N.
Surface Temp. C.	28°.8	28°.8	28°.2	27°.8	27°.3	26°.8	26°.1	26°.2	26°.3	25°.8	20°.3	17°.8	20°.8	22°.7	17°.9
ISOTHERM OF 25° / 77° F.	100	90	105	95	75	75	70	80	50	15	—	—	—	—	—
20° / 68°	130	115	115	105	85	95	110	130	115	120	0	—	15	40	—
15° / 59°	175	130	125	125	115	120	145	185	190	205	205	60	85	105	45
10° / 50°	220	190	160	165	160	195	225	255	270	315	300	150	170	195	145
5° / 41°	600	550	500	600	490	400	430	430	440	475	440	280	325	360	345
2°.5 / 36°.5	1150	900	825	950	860	850	830	900	800	800	720	550	—	—	—
Bottom Temp.	2°.0	1°.0	1°.0	1°.2	1°.3	1°.0	1°.1	1°.0	1°.0	1°.0	1°.2	0°.6	1°.4	3°.3	5°.0
Depth in Fms.	1300	2650	2450	2325	1850	4575	2300	2475	2450	2500	2425	2250	2675	565	345

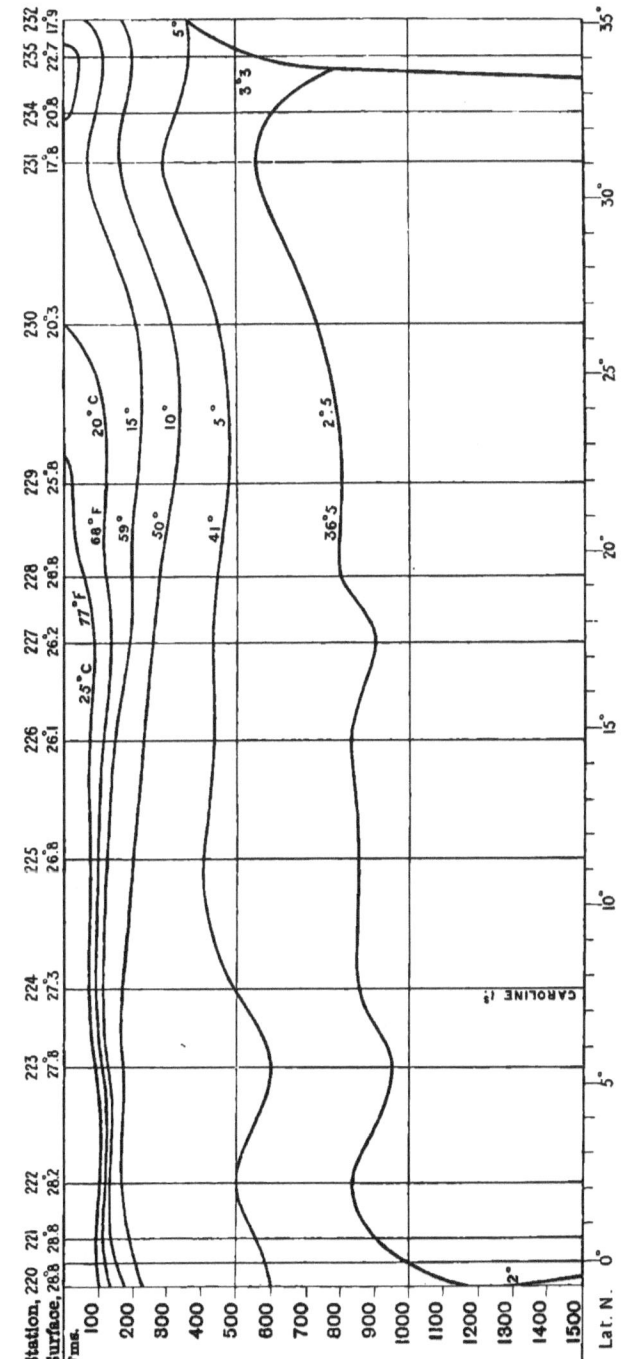

Islands and of the Philippines, and in honour of the first European who crossed the Pacific Ocean. The principal deep-sea communication between this basin and the 3000-fathom area to the eastward is in the narrow sea which flows between the Caroline Islands and the Mariana Islands, where H.M.S. "Challenger" obtained her deepest sounding in 4575 fathoms. The third part of this section, from Station 228 to Station 232, embraces the northern half of the Sea of Magallanes, and is the scene of the encounter between the North Pacific Equatorial Current, here assuming the name of Kuro-Siwo or Japanese Current, and the Arctic Current from the Sea of Okhotsk and the Behring Sea.

One of the most prominent features of this section is the extensive surface-stratum of warm water of a temperature between 29° C. and 25° C. (84° F. and 77° F.), and of a thickness of from 70 fathoms to 100 fathoms. This stratum, which is evidence of a vast accumulation of warm water in the western Pacific, is seen to commence at Station 216 with a depth of 75 fathoms, increase to 100 and 105 fathoms at Stations 220 and 222 in the axis of the Sea of Papua, fall to 75 fathoms and 70 fathoms in the southern part of the Sea of Magallanes, and after gradually thinning off to 50 and 15 fathoms at Station 228 and 229, to disappear altogether. We now enter the waters of the Arctic Current which comes to the surface at Station 231, but from Station 234 to Station 235 we once more find ourselves in a warm surface-stratum, the northern continuation of the North Pacific Equatorial Current known as the Kuro-Siwo, and which is nothing but the Gulf Stream of the North Pacific Ocean, the Sea of Magallanes being on a larger scale the equivalent of the Gulf of Mexico.

No two natural phenomena could present a more complete parallelism than that which can be traced between the origin, progress, and ultimate fate of the great thermal currents of the

North Atlantic and North Pacific Oceans. It constitutes one of the most remarkable proofs of the uniformity of laws and conditions which determine the movements of the oceanic waters from pole to pole. The pouring in of the North Pacific Equatorial Current through the chain of islands which separates the Sea of Magallanes from the main basin of the Pacific, just as the North Atlantic Equatorial Current flows into the Caribbean Sea through the Antilles—the progress of the Pacific current through the southern portion of the Sea of Magallanes, and the accumulation of its waters in the northern and more restricted portion of this sea, as we observe the circulation of the Atlantic current through the Caribbean Sea and the accumulation of its waters in the Gulf of Mexico—the relief of the pressure caused by this accumulation, through the formation in both cases of a powerful current which forces its way through the northern end of the barrier of islands and joins the branch of the equatorial current which has been moving northwards outside this barrier—finally, the subdivision of both equatorial currents, after their encounter with the polar currents, into branches, some of which continue their course into the polar seas, while others bend round, and, gradually cooling in contact with the currents from the north, flow down the western coasts of the opposite continents in order to resume once more their course in the character of equatorial currents, form two parallel series of occurrences, the resemblance between which is too close to be the result of mere accident. An exception to this comparison may be found in the return southwards of a portion of the North Pacific Equatorial Current through the Western Carolines and the Pelew Islands into the Sea of Papua in conjunction with the polar under-current. It is in this latter current that we must seek the cause of the remarkably rapid decrease of temperature in the stratum between 100 and 200 fathoms which forms another prominent feature of this section. The polar current, after its encounter with the equatorial current between Station 234 and Station 229, continues its course as an undercurrent through the Sea of Magallanes. The decrease of temperature in the stratum below 100 fathoms, where the two currents, one flowing south, the other north, are in contact, amounts to about 15° C. (27° F.) in less than a hundred fathoms. A portion of the Arctic current passes down between the Pelew Islands and the Philippines, and we trace its presence in the high level of the isotherms of 5° and 2°.5 C. between Station 218 and Station 214.

A comparison of the isotherms between the latter stations with those of the Sea of Celebes, of the Molucca Passage, the Arafura Sea, and the observations made by the "Gazelle" between North-West Australia and Timor, leaves little doubt but that the Arctic current, after sending a branch into the Sea of Celebes, flows as an under-current through the above-described deep channel or "fault" between the plateau of the Indian Archipelago and the Papua-Australian plateau, for along the whole length of this channel we find the isotherms of 5° C. and 2°.5 C. at about the same depth, the former in an average depth of 500 fathoms, the latter in an average depth of 900 fathoms. A branch of this current probably flows through the Straits of Manipa, past Amboina, into the Banda Sea, and out of the latter past Timor into the Indian Ocean. The isotherms of the China Sea show that the Arctic current also finds its way into that basin. Another branch of the Arctic under-current turns eastward, and, in conjunction with the southern branch of the North Pacific Equatorial Current flowing from the Sea of Magallanes into the Sea of Papua, is the probable cause of the sinking of the isotherms of 5° C. and 2°.5 C. between Station 224 and Station 218, or between the Papuan plateau from Humboldt Bay to the Admiralty Islands and the Caroline Islands (Plates 16 and 17, and Curve Fig. 10).

TABLE XII.—TEMPERATURES OBSERVED IN THE NORTH PACIFIC, BETWEEN YOKOHAMA AND STATION No. 253—*June, July, 1875.*

Station No.	Latitude and Longitude	Surface Temp. C. / F.	25° / 77°	20° / 68°	15° / 59°	10° / 50°	5° / 41°	2°.5 / 36°.5	Bottom Temp.	Depth in Fms.
236	34° 58' N., 139° 30' E.	19.2	—	—	70	185	400	800	2.8	775
237	34° 37' N., 140° 32' E.	22.8	—	50	140	225	390	700	1.7	1875
238	35° 18' N., 144° 8' E.	21.4	—	20	130	240	465	1050	1.0	3950
239	35° 18' N., 147° 9' E.	21.2	—	20	150	255	360	900	1.1	3625
240	35° 20' N., 153° 39' E.	18.2	—	—	20	60	150	600	1.0	2900
241	35° 41' N., 157° 42' E.	20.7	—	15	110	210	340	725	1.1	2300
242	35° 29' N., 161° 15' E.	20.3	—	10	60	170	320	665	1.1	2575
243	35° 24' N., 166° 35' E.	21.7	—	20	75	245	350	750	1.0	2800
244	35° 22' N., 169° 53' E.	21.4	—	10	50	200	345	800	1.2	2900
245	36° 23' N., 174° 31' E.	20.6	—	5	25	190	340	670	1.0	2775
246	36° 10' N., 178° 0' E.	22.7	—	10	40	245	415	800	1.3	2050
247	35° 49' N., 179° 57' W.	22.8	—	20	40	205	380	770	1.2	2530
248	37° 41' N., 177° 4' W.	20.7	—	5	25	140	300	700	1.1	2900
249	37° 59' N., 171° 48' W.	18.4	—	—	25	160	370	710	1.0	3000
250	37° 49' N., 166° 47' W.	18.3	—	—	20	175	430	770	1.0	3050
251	37° 37' N., 163° 26' W.	18.3	—	—	20	155	345	720	1.1	2950
252	37° 52' N., 160° 17' W.	18.3	—	—	20	130	300	700	1.1	2740
253	38° 9' N., 156° 25' W.	19.8	—	—	20	100	300	680	1.0	3125

TEMPERATURES IN THE NORTH PACIFIC,
BETWEEN
YOKOHAMA AND STATION No. 253.
JUNE, JULY, 1876.

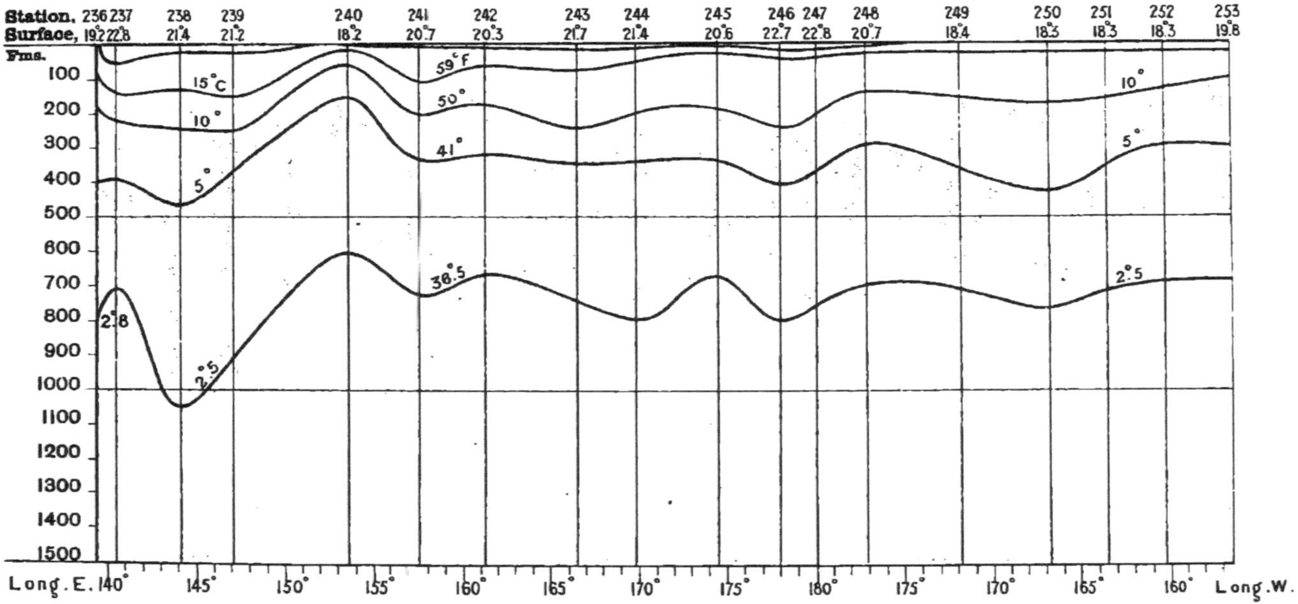

Between Station 225 and 229 we observe the usual fan-like arrangement of the isotherms, caused by the sinking of the heavier equatorial water through the lighter strata of the polar current (Plates 9 and 19). The Kuro-Siwo, running at Station 234 and Station 235, at a short distance from the south coast of Nipon, flows, like the Gulf Stream, over and between the cold waters of the Arctic current, contending with the latter for the alternate possession of the romantic bays and inlets of the south coast of Nipon, Sikok, and Kiusiu. A branch of the Kuro-Siwo penetrates through the Straits of Korea into the Sea of Japan.

The usual alternation of streaks of warm and of cold water which characterise the scene of the meeting between equatorial and polar currents was observed by the "Challenger" during the last three days of her cruise to Japan. Between midnight of the 8th and the morning of the 11th April, 1875, the ship crossed three streaks of cold water of a surface-temperature of 17° C., divided from each other by warm streaks of a temperature of 20° C. The most northern streak entered the Bay of Yokohama, falling to a temperature of 13°.3 C. at the latter port. During the stay of the expedition in Japan, while the water of the Kuro-Siwo outside ranged from 20° to 23° C., the surface-temperature of the bays and inlets of the south coast of Nipon varied between 15° and 17° C.

SECTION FROM YOKOHAMA TO STATION 253 (Plate 18, Table XII.).—This section crosses the North Pacific Ocean between the parallels of lat. 34° N. and lat. 38° N., from the coast of Japan to the meridian of the Sandwich or Hawaiian Islands. Its western portion exhibits the relations between the equatorial and the polar currents eastward of Nipon. After traversing the belt of cold water which fringes the east and south coast of this island, we enter, at Station 237, the Kuro-Siwo at the point where it joins the main stream of the North Pacific Equatorial Current, which flows outside the line of islands that separate the northern part of the Sea of Magallanes from the North Pacific basin. Like the Gulf Stream, the Kuro-Siwo imposes its name upon its more powerful though less conspicuous parent. The axis of the current is at Station 238, where it depresses the isotherm of $2°.5$ C. from its average North Pacific level at 700 fathoms down to below 1000 fathoms. This axis corresponds with the axis of the 4000-fathom channel, which, as formerly described, stretches northward along the coast of Nipon and Yezo. The breadth of the warm current, measured from its western limit off the Japanese coast to beyond Station 239, is over 400 miles. At Station 240 we find ourselves in the middle of a great polar current which flows down between Station 239 and Station 241 in a south-westerly direction, and reduces the temperature of the water to a depth of more than 600 fathoms. This is the same current whose course we have been tracing through the Sea of Magallanes, past the Pelew Islands, into the Molucca Passage, and through the Indian Archipelago into the Indian Ocean. It probably divides itself into two branches, one entering the Sea of Magallanes north of the Bonin Islands, and between the Bonin and the Mariana Islands (Stations 228–231), and continuing its south-westerly course towards the Philippines, the other turning down outside these islands into the 3000-fathom basin situated north of the Carolines.

The isotherms of the stations to the eastward of Station 240 indicate the existence of alternate warm and cold currents—the former, branches of the equatorial current flowing first eastward, then turning southward, across the parallel of lat. 40° N.; the latter, cold currents from the sea of Okhotsk and the Behring Sea. There are warm currents at Stations 241, 243, 246, and 250, cold currents at Station 242, between 244 and 245, and at Station 248. A cold current seems to flow down on each side of the projecting north-western extremity of the Hawaiian plateau at Station 246. From Stations 248 to 253, after cross-

From Station 253 to Station 288. 117

ing the meridian of long. 180°, we pass into the thermal area of the North-Eastern Pacific.

SECTION FROM STATION 253, ALONG THE MERIDIAN OF HONOLULU AND TAHITI, TO STATION 288 (Plate 19, Table XIII.).— Embracing nearly 80 degrees of latitude, and extending along a track of considerably over 5000 nautical miles divided into 35 stations, this section, surveyed in the third year of the "Challenger" cruise round the world, is a lasting monument of the skill and perseverance of the officers and men of the old English frigate.

A minute examination of the section could only lead to an unnecessary repetition of much that has been said in connection with the other sections. With the assistance of the sketch given in previous chapters of the leading phenomena of oceanic circulation, it will not be difficult to arrive at the principal facts connected with the distribution of temperature in the Pacific Ocean, viz. : The warm surface-stratum between the parallels of lat. 30° N. and lat. 30° S., the "cold wall" between the 35th and 40th parallels, and the gradual warming of the intermediate strata indicated by the spreading out of the isotherms from the equatorial belt towards the 35th parallel.

A comparison of the Atlantic section (Plate 9) with the Pacific section (Plate 19) brings out the principal contrast between the two oceans. While the North Atlantic basin is considerably warmer than the South Atlantic basin, we observe the contrary in the Pacific Ocean ; or it would be more correct to say that the South Pacific is warmer than the South Atlantic, and the North Pacific colder than the North Atlantic, since the two sections do not afford a fair comparison between north and south in the two oceans. The differences observed between the Atlantic and the Pacific are due chiefly to the great difference between their respective areas. Owing partly to the projection of the South American coast at Cape S. Roque into the com-

TABLE XIII.—TEMPERATURES OBSERVED IN THE PACIFIC OCEAN, BETWEEN LAT. 40° N. AND LAT. 40° S., OR BETWEEN STATION NO. 253 AND STATION NO. 288—*July to October, 1875.*

Station No.	Latitude and Longitude	Surface Temp.	\[Isotherm of F 77° / C 25°\]	F 68° / C 20°	F 59° / C 15°	F 50° / C 10°	F 41° / C 5°	F 36.5° / C 2.5°	Bottom Temp.	Depth in Fms.
288	40° 3′ S, 132° 58′ W	12°.5	—	—	—	95	465	900	0°.8	2600
287	36° 32′ S, 132° 52′ W	14°.3	—	—	—	145	475	850	0°.8	2400
286	33° 29′ S, 133° 22′ W	17°.2	—	—	110	185	500	950	0°.8	2335
285	32° 36′ S, 137° 43′ W	18°.3	—	—	135	225	500	900	1°.0	2375
284	28° 22′ S, 141° 22′ W	20°.0	—	0	150	250	475	825	1°.0	1985
283	26° 9′ S, 145° 17′ W	20°.3	—	30	160	240	450	870	1°.3	2075
282	23° 46′ S, 149° 59′ W	22°.9	—	90	190	250	500	840	1°.0	2450
281	22° 21′ S, 150° 17′ W	23°.6	—	105	180	250	450	900	0°.8	2385
280	18° 40′ S, 149° 59′ W	25°.1	0	135	195	240	445	830	1°.2	1940
279	Off Papeete, Tahiti	26°.1	35	140	180	240	—	—	—	680
278	17° 12′ S, 149° 43′ W	26°.4	55	140	195	235	460	850	1°.6	1525
277	15° 51′ S, 149° 41′ W	26°.1	40	140	190	—	—	—	1°.0	2325
276	13° 28′ S, 149° 30′ W	26°.7	75	135	175	225	430	940	1°.0	2350
275	11° 20′ S, 150° 30′ W	26°.7	65	120	160	220	500	870	0°.9	2610
274	7° 25′ S, 152° 15′ W	26°.8	75	105	125	165	500	950	0°.9	2750
273	5° 11′ S, 152° 56′ W	27°.0	75	100	120	175	435	900	0°.7	2350
272	3° 48′ S, 152° 56′ W	26°.1	60	90	115	180	450	860	1°.0	2600
271	0° 33′ S, 151° 34′ W	26°.0	10	80	100	210	500	930	1°.0	2425
270	2° 34′ N, 149° 6′ W	26°.4	30	90	110	290	700	1130	0°.7	2925
269	5° 54′ N, 147° 2′ W	27°.3	70	85	100	150	500	930	1°.1	2550
268	7° 35′ N, 149° 49′ W	27°.2	45	55	120	500	1060	—	0°.8	2900
267	6° 28′ N, 150° 49′ W	26°.7	10	35	45	130	470	900	0°.8	2700
266	11° 7′ N, 152° 3′ W	26°.7	15	45	60	155	450	780	1°.0	2750
265	12° 42′ N, 152° 1′ W	26°.2	25	55	140	550	860	—	0°.8	2900
264	14° 19′ N, 152° 37′ W	25°.3	40	60	85	150	470	960	1°.2	3000
263	17° 33′ N, 153° 36′ W	25°.3	30	60	90	150	500	1040	—	2650
262	19° 12′ N, 154° 14′ W	25°.3	30	80	105	145	500	820	1°.2	2875
261	20° 18′ N, 157° 14′ W	25°.8	5	60	105	150	300	890	1°.3	2050
260	21° 11′ N, 157° 25′ W	24°.9	—	70	100	160	—	—	6°.7	310
259	23° 3′ N, 156° 9′ W	25°.0	0	80	125	180	370	930	1°.0	2225
258	26° 11′ N, 155° 12′ W	25°.0	0	45	100	175	350	760	—	2775
257	27° 33′ N, 154° 55′ W	24°.7	—	35	175	330	800	—	1°.0	2875
256	30° 22′ N, 154° 56′ W	23°.3	—	35	85	170	330	730	1°.2	2950
255	32° 28′ N, 154° 33′ W	23°.3	—	20	50	160	345	740	1°.0	2850
254	35° 13′ N, 154° 43′ W	22°.2	—	15	35	130	350	800	1°.0	3025
253	38° 9′ N, 156° 25′ W	19°.8	—	—	20	100	300	680	1°.0	3125

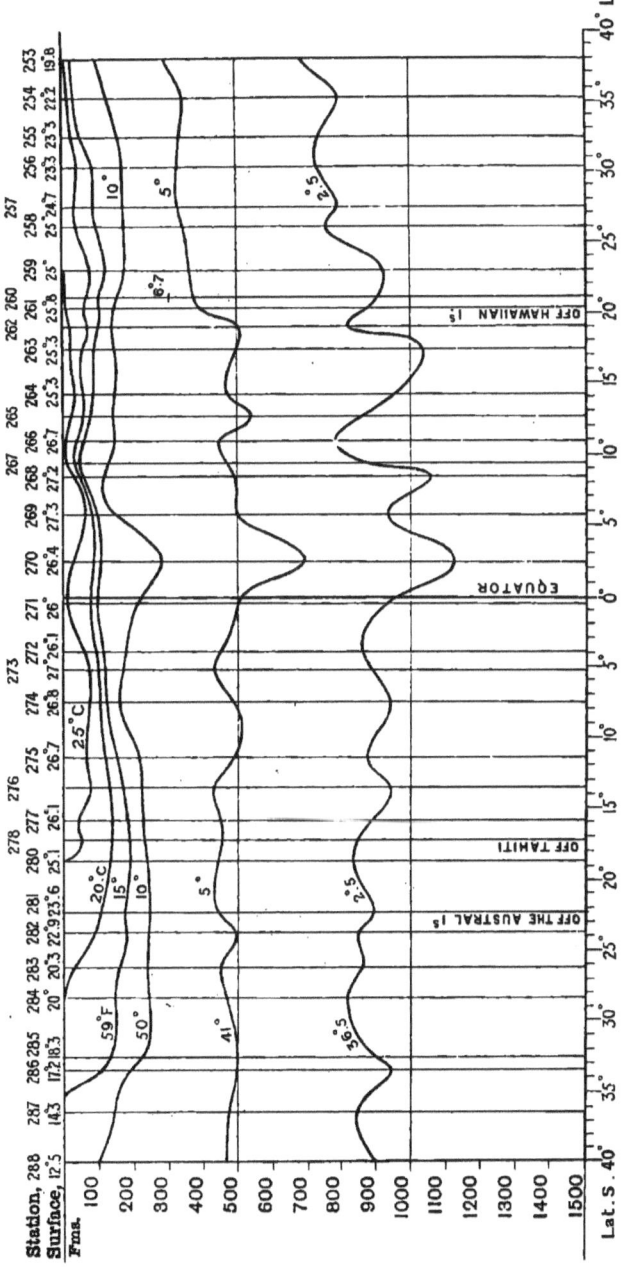

paratively narrow channel between the latter cape and Cape Palmas in Africa, and partly to the fact that the northern limit of the south-east trades is placed for the greater part of the year north of the equator, a considerable portion of the South Atlantic Equatorial Current is diverted along the north-east coast of South America into the Northern Atlantic. On the other hand, nearly the whole of the South Pacific Equatorial Current remains on its own side of the equator, hence the western half of the South Pacific is much warmer than the western half of the South Atlantic. Although the North Atlantic receives such an important contribution from the South Atlantic, the accumulation of warm water in its western half is much inferior to that in the western half of the North Pacific; yet, on account of the restricted area of the Atlantic, its eastern half, both north and south of the equator, receives a greater quantity of warm water in the shape of equatorial return-currents than the eastern half of the Pacific, and the eastern half of the former is therefore much warmer than that of the latter.

Another remarkable phenomenon in the circulation of the oceanic waters, and one the explanation of which has exercised many minds, is to be found in the equatorial counter-currents of the Atlantic, Indian, and Pacific Oceans. In all the three oceanic basins this current is identified with the equatorial belt of calms, and is known to flow from west to east, while the equatorial currents on both sides of it flow from east to west. Its presence in the belt of calms, to which it is exclusively confined, is no doubt not accidental, and leads to the conclusion that the absence of all permanent atmospheric currents must be a circumstance favourable to the formation of a current with a tendency, as in the present case, to flow in an easterly direction. At the same time, we are led to examine whether there is anything in the temperature-conditions of this belt and of the zones immediately adjoining which may cause the formation of a current flowing from west to east. The zones in the immediate vicinity of the belt of calms are occupied by the equatorial currents, the surface-stratum of which, under the influence of the trade-winds, moves from east to west. At the same temperature, or at nearly equal temperatures, the surface-stratum of these currents must be of less specific gravity than the surface-water of the belt of calms, and this difference would cause the water of the equatorial currents to flow over at the limits of the belt of calms where the trade-winds are no longer felt. At the same time, this difference of specific gravity would be greatest near the eastern end of the belt of calms, where the currents arriving from higher latitudes turning to westward assume for the first time the character of equatorial currents; and least towards the western end, where the surface-water of these currents, after flowing for some time under an equatorial sun, has already increased in specific gravity. This circumstance would give the water flowing over into the belt of calms a permanent tendency to flow eastward, which, through accumulation of effect along the whole length of the belt, would establish and maintain a permanent current flowing from west to east in the equatorial belt of calms. It may be remembered that, in the absence of atmospheric currents and of differences of temperature, the specific gravity of the water is the sole arbitrator in the arrangement of strata—a principle which, as we have seen, applies on a much larger scale to the changes which take place in the belt of calms of the parallel of lat. 30° N. and S., and in the polar areas of calms.

Actual observation has established the fact of the overflowing of the water of the equatorial currents into the belt of calms, and, as we might expect, chiefly along the southern limit of the belt. The counter-current, arrested by the continents which stretch across its course, in its turn overflows north and south, and rejoins the equatorial currents on each side of it.

A more simple explanation of this counter-current may be found in the fact that the equatorial currents, as they flow on each side of the belt of calms, remove the water from the eastern and accumulate it at the western side of the basin, and that the counter-current tends to restore the equilibrium thus constantly disturbed.

SECTION FROM STATION 288 TO VALPARAISO AND MAGELLAN STRAITS (Plate 20, Table XIV.).—This section may be compared with the soundings between Tristan d' Acunha and the Cape of Good Hope (Plate 11). It extends for the greater part along the parallel of lat. 40° S. between the meridians of Pitcairn Island and Juan Fernandez, and coincides with the limit between the thermal areas of the South Pacific and of the Southern Ocean. With rapidly-decreasing temperatures in the surface-stratum between lat. 30° and 40° S., we arrive, at Station 288 on the 40th parallel with a surface-temperature of 12°.5 C., and the isotherm of 10° C., at a depth of 95 fathoms. Proceeding eastward, we observe a sudden rise of the latter isotherm from 95 fathoms to 45 fathoms at Station 288, and to 40 fathoms at Station 290, gradually falling to 90 fathoms at Station 294, indicating a cold surface-current from the south between Stations 289 and 292. The increase of temperature further eastward between the surface and 100 fathoms, combined with the lessening latitude of the stations, shows that we enter more and more into the warmer water of the equatorial return-current which flows over the plateau of Juan Fernandez (Plate 5), especially perceptible at Station 296; while the sinking down of the isotherm of 2°.5 C. between Stations 292 and 296 attests the presence of a warm under-current, which, flowing along the southern slope of the plateau, turns southward towards the Antarctic shores of Graham Land. Part of this current is probably carried through the Strait of Cape Hoorn. On the other hand, the rise of the isotherms of 2°.5, 5°, and 10° C. at Stations 297, 301, 302, and

TABLE XIV.—TEMPERATURES OBSERVED IN THE SOUTH PACIFIC, BETWEEN STATION No. 288 AND THE COAST OF CHILE—*October to December, 1875.*

Station No.	288	289	290	291	292	293	294	295	296	297	298	299	300	301	302	303
Latitude and Longitude	40° 3′ S, 132° 58′ W	39° 41′ S, 131° 23′ W	39° 16′ S, 124° 7′ W	39° 13′ S, 118° 49′ W	38° 43′ S, 112° 31′ W	39° 4′ S, 105° 5′ W	39° 22′ S, 98° 46′ W	38° 7′ S, 94° 4′ W	38° 6′ S, 88° 2′ W	37° 29′ S, 83° 7′ W	34° 7′ S, 73° 56′ W	33° 31′ S, 74° 43′ W	33° 42′ S, 78° 18′ W	37° 29′ S, 84° 2′ W	42° 43′ S, 82° 11′ W	45° 31′ S, 78° 9′ W
Surface Temp.	12°.5	12°.5	11°.4	11°.7	11°.8	12°.0	14°.2	14°.7	15°.4	13°.9	15°.0	16°.7	17°.0	15°.3	12°.8	12°.7
Isotherm of 25° C. / 77° F.	—	—	—	—	—	—	—	—	—	—	—	—	—	—	—	—
Isotherm of 20° / 68°	—	—	—	—	—	—	—	—	—	—	—	—	—	—	—	—
Isotherm of 15° / 59°	—	—	—	—	—	—	—	—	10	—	0	25	25	10	—	—
Isotherm of 10° / 50°	95	45	40	60	60	75	90	85	100	100	90	85	100	100	45	40
Isotherm of 5° / 41°	465	465	465	400	400	460	400	375	370	355	400	350	300	400	300	275
Isotherm of 2°.5 / 36°.5	900	850	800	800	830	940	930	870	1050	800	900	820	840	—	840	800
Bottom Temp.	0°.8	0°.8	0°.9	0°.8	1°.3	0°.7	0°.7	1°.4	1°.2	1°.3	1°.3	1°.1	1°.5	—	1°.5	1°.8
Depth in Fns.	2600	2550	2300	2250	1600	2025	2270	1500	1825	1775	2225	2160	1375	—	1450	1325

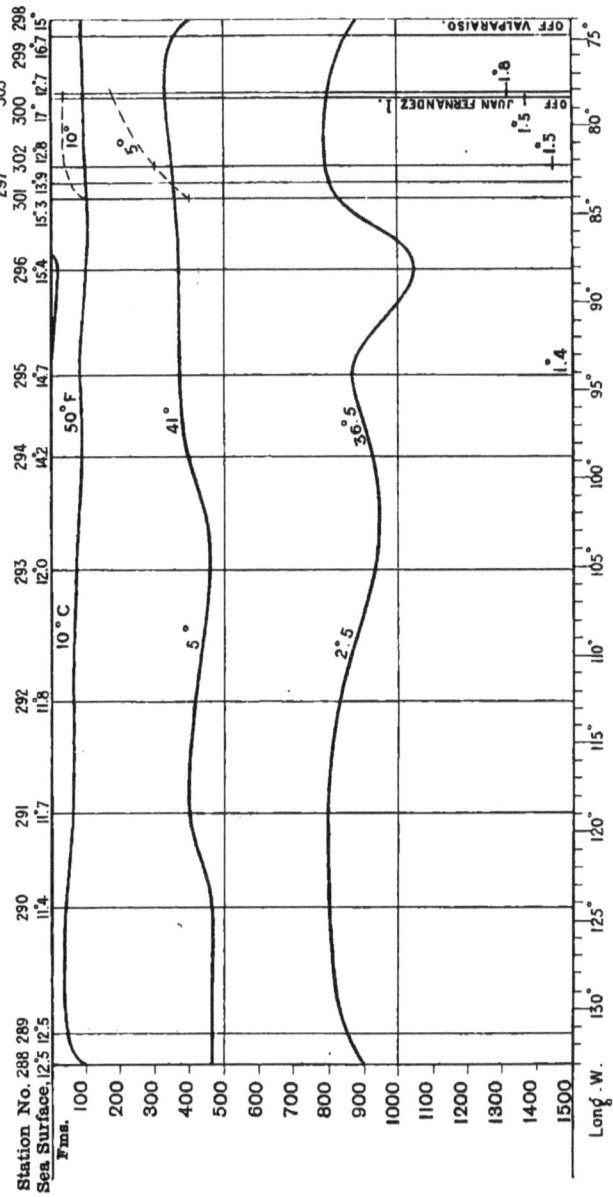

303, shows that on approaching the coast of Patagonia we enter the southern branch of the Antarctic current, which, bending round, flows southward along that coast and through the Strait of Cape Hoorn. Its northern branch sweeps over the plateau of Juan Fernandez, and follows the coast of South America up to the equator (Plate 5).

The Bay of Valparaiso must present contrasts of temperature similar to those observed in False Bay near the Cape of Good Hope. A westerly wind would bring the warm water of the equatorial return-current into the bay, while a southerly wind would fill the latter with the cold water of the Antarctic current. The latter name is perhaps not a correct designation of the great surface-current which, flowing from west to east through the Southern Ocean, makes the circuit of the world between the parallels of lat. 40° and 60° S. and forms the "cold wall," on encountering which all equatorial return-currents are split up into currents running eastward along this wall, and into currents flowing as warm under-currents into the Antarctic region.

CHAPTER VI.

THE BED OF THE OCEAN.

Changes in the Distribution of Land and Water—Formation of Sub-oceanic Strata—Formation of Central Oceanic Plateaux—Formation of Areas of Elevation and of Areas of Depression—Formation and Transformation of Continents—Formation of Mountain Ranges and Submarine Ridges.

CHANGES IN THE DISTRIBUTION OF LAND AND WATER.—It was mentioned in an earlier chapter that the ordinary conception of the relative distribution of land and water over the surface of the earth may be replaced or rather supplemented by one which more adequately embodies the results of modern research, and according to which the surface of the solid earth-crust may be considered as composed of hills and hollows, areas of elevation and areas of depression—the former not necessarily constituting dry land, the latter not always occupied by water. It was also shown how the data furnished by recent sounding operations afford additional evidence of the observation—not made for the first time, since it has attracted the attention of every student of comparative geography—that the principal land-masses, more or less combined into one great area of elevation, gravitate towards the North Pole as their common centre; while the different oceanic basins, constituting one great area of depression, gather round the South Pole as their centre. If this observation conveys any information beyond the familiar fact that there is more land in the northern and more water in the southern hemisphere, it means that the slow but unceasing changes which take place in the distribution of land and water obey a general tendency to accumulate land in the northern and water in the southern hemisphere. There are numerous indica-

Fig. 13.
Diagram showing Decrease of Diamater of Rotation, from the Equator to the Poles._

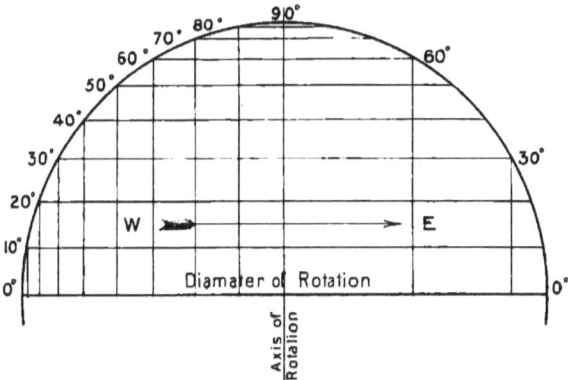

Fig. 14.

Fall of 1 Mile in 5 Miles.

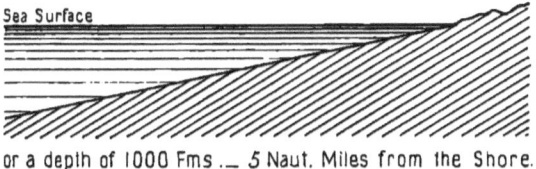

or a depth of 1000 Fms._ 5 Naut. Miles from the Shore.

Fall of 1 Mile in 10 Miles.

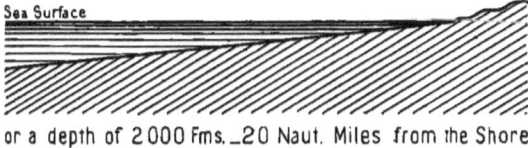

or a depth of 2000 Fms._20 Naut. Miles from the Shore

Fall of 1 Mile in 20 Miles

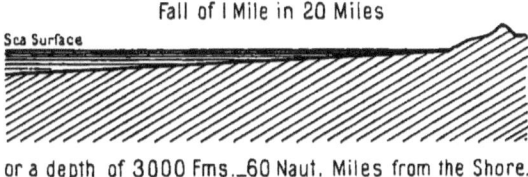

or a depth of 3000 Fms._60 Naut. Miles from the Shore.

tions of a similar tendency to transfer land and water from east to west, so that a combination of both tendencies would result in a general movement of land from south-east to north-west, and of water from north-east to south-west.

The investigation of the problem suggested by this general movement of land and water, if it really exists, seems to belong more to the domain of the astronomer than of the student of physical geography, since the transfer of great masses of solid and fluid matter could not, apparently, take place without affecting the distribution of terrestrial gravity, the position of the axis of rotation, &c. However, instead of invoking cosmic agencies which sometimes escape the grasp of the most accomplished mathematician, it may be possible to discover causes whose action is more within reach of direct observation, and which may afford a sufficient explanation of the phenomenon above alluded to.

At the outset it appears, from a comparison of the height of the protuberances or of the depth of the hollows which compose the surface of the solid earth-crust with their lateral extension, that even a slight elevation or depression of portions of that surface, insignificant in amount when contrasted with the diameter of our planet, may produce a considerable change in the distribution of land and water. According to the soundings taken in every part of the ocean, an elevation or depression amounting to 100 fathoms, the eighty-thousandth part of the earth's diameter, would completely change the outlines of the dry land as they are at present laid down in our charts. Great Britain, for example, would either form part of the Continent of Europe, or be reduced to a cluster of small islands rising out of the sea at a great distance from the French coast, formed by the slopes of the Ardennes, the Vosges, and the mountains of Auvergne. It so happens that both events have occurred in the past. The effect which such a change of level must have upon oceanic and atmospheric currents, upon climate, and upon the whole fauna and flora of the region where it takes place, may be readily appreciated.

The average height of the dry land above the level of the sea has been calculated to amount to less than 200 fathoms, while the average depth of the ocean is probably over 2000 fathoms; so that, if we deduct the mountain ranges and elevated plateaux which largely contribute to the above average, a great portion of the dry land must be less than 100 fathoms above the level of the sea. A depression of 100 fathoms, while it would cause almost all dry land to disappear—all but the most elevated regions—would reduce the depth of the ocean by only one-twentieth. In connection with this subject, it is necessary to guard against an impression produced by recent discoveries of extensive areas of great depth in the vicinity of the land, and encouraged by the small scale on which the results of sounding operations have to be presented to the eye. The comparatively rapid increase of depth, so frequently observed beyond the 100-fathom line, has suggested the idea that the continents of the old and new world rise *abruptly* from the bottom of the sea and form high plateaux, whose steep sides descend within a short distance of the shore into depths of two or three and occasionally four or five miles. It is but natural that a distance of five, ten, or twenty miles should appear very short when compared with the wide expanse of an oceanic basin; but it will become evident, to any one who will take the trouble to put down on paper the proportion between distance and depth, that a depth of 1000 fathoms, or of one mile, at a distance of five miles from the shore, by no means forms what is generally understood by a "steep incline," as the angle is little over 11°. A depth of one mile at a distance of ten miles is a comparatively rare occurrence, and in most cases where the soundings seem to increase with more than usual rapidity to depths of 2000 and

3000 fathoms, the distance from the shore at which they are found is seldom less than forty or sixty miles—that is to say, a descent of one mile in twenty miles, or an angle of about 3° (Fig. 14). On measuring the inclines of several islands of volcanic origin, such as Pico in the Azores, Ascension Island, Marion Island, and the island of Hawaii, as these appear on sketches made during the cruise of H.M.S. "Challenger," the angle is found to decrease from an average of 30° at the crater or craters, to 15° and 10° upon the intermediate slopes, while the final incline dips into the sea at an angle of from 10° to 6°—that is to say, a fall of one mile in ten miles, which a few miles from the shore is reduced to 3°, or a fall of one mile in twenty miles. Yet those islands have the aspect of rising abruptly from the level of the sea, and depths of over 2000 fathoms are obtained within a few hours' sail from their shores.

The purpose of the above remarks is to point out that continents are but the most elevated areas of wide and low undulations, and that these differ in no respect from the submarine plateaux discovered by recent exploring expeditions, except in having partially risen above the surface of the ocean. We also see that, on account of the low angle of the inclines, a comparatively slight alteration either in the level of the land or in the level of the sea may produce a considerable change in the distribution of land and water, and that the rise and fall of these undulations rarely exceed five miles in a distance of 100 miles, and are generally much below this proportion.

The comparatively rapid increase of depth beyond the 100-fathom line was a phenomenon of sufficiently frequent occurrence to attract the attention of those engaged in the recent sounding operations, and can hardly be considered as accidental. It has probably some connection with the limits of the alterations of level which have taken place during the most recent geological period, and which apparently do not range beyond the 100-fathom line.

FORMATION OF SUB-OCEANIC STRATA.—Oscar Peschel, in his remarkable essay on *New Problems in Comparative Geography*, has already expressed the opinion that the continents are older than the mountain ranges we find upon them; that the latter have been raised up along the coast-lines of the former, and that their elevation appears to be due to lateral pressure. He also remarks that most of these coast-ranges are backed on the land side by high plateaux. A study of the results of recent deep-sea exploration will lead to the same conclusions.

If there be little doubt that the currents of the ocean are the carriers and distributors of temperature throughout the vast depths of the seas which cover so large a portion of the surface of our planet, it is equally clear that water in every shape, from the smallest stream to the great oceanic rivers, is the principal solvent, carrier, and distributor of the solid matter which composes the only portion of the earth-crust with which we are acquainted. The solid particles thus held in suspense are deposited according to their weight and bulk—the heavier ones first and nearest to the place whence they came, whilst the lighter ones are carried to a greater distance. Those which are light and yet bulky remain in suspense for some time: if lighter than water, they will never reach the bottom; if a little heavier, they will do so only after a lapse of time, longer or shorter according as the conditions are more or less favourable. Chief amongst these conditions is the velocity of the current of water which acts as the carrier of solid matter; the greater that velocity, the greater is the weight of the solid particles held in suspense, and the greater is the distance to which they are carried, and *vice versa*.

The matter distributed by oceanic currents is mainly composed of inorganic detritus, the result of sub-aërial and submarine denudation, of organic remains derived from plants and animals, and of substances held in solution, such as salts, gases, &c. In accordance with the above-mentioned conditions, we may expect the solid particles to form deposits varying in quantity and quality in proportion to the distance to which they have been carried, and to the greater or lesser velocity of the currents which occupy the area in which they have been deposited. This conclusion is borne out by facts which have come to light in the course of the recent researches into the nature and composition of the deposits found at the bottom of the sea.

The samples brought up from the bottom in the tube of the sounding apparatus reveal a marked difference between deposits formed near the land and deposits accumulated in the more central parts of an oceanic basin. This difference is sufficiently great to render it possible—as soon as we shall possess a complete analysis of the specimens already collected—to decide whether a certain sample of the sea-bottom, the origin of which may be doubtful, belongs to a stratum deposited in a deep sea or in a shallow sea, near the margin or near the centre of an oceanic basin.

It is evident that a large proportion of the detritus derived from sub-aërial and submarine denudation, including all the heavier and at the same time more voluminous particles, will be deposited within a short distance from the margin; that the composition of this marginal deposit will depend upon, and vary with, the materials which make up the surface-strata of the adjoining land; and that the distance to which it extends from the shore will be influenced by the presence or absence of shore-currents, or of rivers emptying themselves into the sea. Thus the breadth of the marginal deposits may amount to several hundred miles at the mouth of great rivers, such as the Amazon, the Rio de la Plata, the Mississippi, &c., while it may be reduced to a few miles in places where the shore is swept by powerful currents. Hence, under certain conditions, large submarine plateaux may be formed in connection with the land; their rate of accumulation will be comparatively rapid, and they will have a tendency to alter the configuration of the basin as well as the direction of its currents.

The lighter and finer particles are carried to a greater distance from the margin, and deposited in the more central parts of the basin. Their rate of accumulation will be much slower; so that an oceanic as well as an inland basin has a tendency to fill up from the margin towards the centre, and may end in being completely filled up, unless this gradual accumulation is kept in check through the action of currents which remove a portion of the deposits and transfer it elsewhere. This transference is the general rule, for as the bed of a current becomes more and more restricted, its velocity increases in the same proportion, and with it its power to remove part of the deposits. The distance to which the lighter and finer particles are carried by oceanic currents before they arrive at their final resting-place may amount to several thousand miles, and this is probably the cause of the remarkable uniformity which has been observed in the character and composition of the deposits formed not only over vast areas of the same basin, but also in the different oceanic basins, as compared with the variety which exists in the composition of marginal deposits.

Hence we may infer that a stratum of a nearly uniform character, which is found to extend over wide areas, must have been deposited at the bottom and towards the centre of an oceanic basin; while strata of lesser extent, and offering a greater variety in their composition, must have been marginal deposits.

FORMATION OF CENTRAL OCEANIC PLATEAUX.—The lighter and finer particles distributed by oceanic currents may be divided, according to their origin, into inorganic and organic particles. The former are, as a rule, much heavier in comparison with their volume than the latter, and will therefore be deposited sooner than organic particles, which can only fall to the bottom and form strata under peculiarly favourable conditions. The most favourable of these conditions is the absence or nearly complete absence of currents, and this conclusion is remarkably confirmed by observation. A large proportion, if not by far the largest proportion, of the particles suspended in the waters of the ocean consists of the bodies of the myriads of animal and plant organisms which there live and die, and no doubt derive their sustenance from the still finer organic and inorganic particles dissolved in the surrounding fluid. A teaspoonful of salt water examined under the microscope reveals the presence of hundreds and thousands of these minute organisms. Their distribution in the ocean, no less than that of the larger animals, depends, among other conditions, upon the nature and abundance of the food they require; and hence we can distinguish between a marginal and a central oceanic fauna, and between a surface and a bottom fauna. Although the dredge has brought to light sufficient proofs of the presence of animal life at great depths—two or three miles from the surface—yet a considerable diminution has been observed beyond these limits, ending with an almost complete absence of living organisms as we attain a depth of 4000 fathoms. Most of the minute organisms seem to have their home in the upper strata, being especially abundant at or near the surface, and their bodies reach the bottom only after death, and after having floated for a considerable time with the currents. During this time they undergo a process of decomposition which reduces them to a mere skeleton, and the latter, being heavier in proportion to its bulk than the living body, ultimately sinks to the bottom.

The deposit of these light remains of organic life can only take place, as already mentioned, over areas of minimum circulation, and these areas are confined to the centre of oceanic basins and to what we have termed the critical latitudes. We may therefore expect the formation of deposits composed mainly of organic particles in the centre of oceanic basins and in the critical latitudes, where, as is the case with atmospheric currents, we find areas of calms. This conclusion is in harmony with observed facts. The central plateaux of the North Atlantic and of the South Atlantic, the wide plateaux between the latter, the Indian and the South Pacific Ocean, and the Southern Ocean, are all found to be covered with a stratum composed of the remains of the minute organisms which live in the ocean. On the contrary, the bottom of the areas of depression consists principally of inorganic particles in the shape of extremely fine and very tenacious clays, varying in colour from grey to yellow and red, and occasionally deepening to a chocolate colour.

Awaiting a more complete analysis of the specimens of bottom brought up by the sounding-tube in the course of the "Challenger" expedition, the marginal deposits have been described as mud, sand, stones, rock, shells, &c., the deposits in the areas of depression as red clay or grey ooze, and those in the areas of elevation as globigerina ooze. The clays are composed of very fine mineral particles mixed up with a small percentage of organic remains. As the depth decreases, this percentage is found to increase, until, at depths of less than 2000 fathoms, the deposit is almost exclusively composed of the skeletons and fragments of skeletons of the minute forms of animal life which inhabit the ocean.

The previous arguments, which may be modified by future and more detailed research, lead to several conclusions of some importance to the geologist. As the discoveries made by the expeditions on board H.M.S. "Lightning" and "Porcupine," in the area between the Færoe Islands and Scotland, have shown that an Arctic and a southern fauna may exist side by side at a distance of a few miles from each other, so the results of the "Challenger" expedition tend to prove *that the paucity or total absence of organic remains in a geological stratum is no evidence of its relative antiquity.* The difference which is observed between the deposits found in areas of depression and those accumulated in areas of elevation, shows that a comparatively rapid accumulation of organic remains may take place in one portion of an oceanic basin, *contemporaneously* with the slower deposit of a formation which is almost or nearly destitute of organic remains in another portion of the same basin. This remark may be extended to the remains of the higher forms of animal life. Some astonishment was created on board the "Challenger" that the dredge, after having been dragged over miles and miles of the bottom of the sea, and up and down almost every oceanic basin, should never bring up any bones of fish or whale, or any remains of other large animals which inhabit the sea, or whose bodies may have been carried down to the sea; for, with the exception of a few shark's teeth and some ear-bones of whales, no portion of one of the more highly organised animals was ever found in the dredge or in the bag of the trawl, always excepting those forms which we had learned to associate with the bottom of the sea, and which have also been found in abundance in the strata of former geological periods. What becomes of the wrecks innumerable and of the bones of the multitudes who have, in the service of their country and their race, found an honourable grave in the depths of the sea? No portion of a ship or any other article of human manufacture, no human bones, ever came to the surface; and though a satisfactory explanation of this curious fact may yet be found, it shows that we should hesitate before accepting the absence of these remains as conclusive evidence of the antiquity of a geological stratum, or of the non-existence of higher organisms, including man, in former periods.

FORMATION OF AREAS OF ELEVATION AND OF AREAS OF DEPRESSION.—Assuming the previous deductions to be correct, we can now conceive the formation of areas of elevation and areas of depression, in consequence of the unequal distribution of solid matter by oceanic currents, even in the absence of pre-existing dry land, and antecedent to all other phenomena of elevation and depression due to volcanic or other agencies. If we suppose the whole surface of our planet covered with water, the more rapid accumulation of solid matter in areas where there is little or no current, and its slow deposit in areas where strong currents prevail, would after a time divide the bed of the ocean into plateaux and depressions. It is remarkable that, while the direction of the principal mountain ranges is so frequently from north to south, the direction of the lines which divide the great river systems of our continents is generally from east to west, which would imply that the longitudinal axis or central ridge of the original plateaux, previous to the rising up of the mountain ranges, was from east to west (Plate 4A). The conversion of submarine plateaux into dry land would be effected by the gradual rising of the areas of elevation through continuous accumulation of solid matter, simultaneously with the deepening of the areas of depression through the removal of deposits by currents, whose velocity must increase as their area becomes more restricted. It may also be the result of a diminution in the total quantity of water contained in the ocean, or of a retreat of the ocean, for there seems to be no argument to prove that this quantity must be constant, or that the level of the ocean must always be exactly at the same distance from the centre of terrestrial gravity. On the contrary, if we separate the centre of gravity of the whole mass of oceanic waters from the centre of gravity of the solid portion of our planet, the former may be subject to certain fluctuations, as the latter must be affected by changes in the arrangement of the solid earth-crust.

Both may be said to move round their common centre of gravity, which is that of the whole planet, and, in consequence, we may conceive a gravitation of the whole mass of the ocean in one direction—for example, in favour of the southern hemisphere, which would result in the sinking of the level of the ocean—*i.e.*, the creation of dry land in the northern hemisphere. The great plains of the present continents, such as the plain which stretches from the English Channel to the coast of Siberia, and the great North American plain, have all the appearance of having been converted into dry land, not through the action of a subterraneous agency which lifted them above the level of the sea, of which action they bear little or no trace, but in consequence of the retreat of the ocean; and the great rivers which now wind their course through these plains have carved out their bed, not through strata previously deposited by themselves in a former geological epoch, but through strata deposited at the bottom of former oceanic basins and great inland seas.

Or, supposing the quantity of oceanic waters to be constant, the bed of the ocean may be either deepened or rendered more shallow, or its total area made wider or narrower, in consequence of submarine denudation, the formation of new seas, the accumulation of fresh deposits, and the uplifting or depression of wide areas by subterraneous forces. Any such alteration in the contour of its bed, of the occurrence of which in the past and in the present there is ample evidence, may either raise or lower the level of its surface, and it has already been shown how profoundly the distribution of land and water would be affected by a change of level amounting to the merest fraction of the total depth of the ocean.

FORMATION AND TRANSFORMATION OF CONTINENTS.—If we suppose that at one time the ocean covered the whole surface of the earth, the plateaux accumulated in consequence of the unequal distribution of solid matter by the thermal oceanic currents would occupy the critical latitudes, and their direction would be from east to west, or parallel with the equator. We should have an equatorial plateau separated by zones of depression from plateaux occupying the parallels between lat. 30° and 50° N. and S., which again would be divided by zones of depression from the plateaux of the polar regions, and the surface of our planet would have the appearance of being divided into more or less parallel strips composed of alternate areas of elevation and depression.

The elevation of ridges parallel with the axis of these plateaux, and due to what at present is termed volcanic or subterraneous agency, would at once cause a change in the system of oceanic circulation, and consequently in the distribution of solid matter. As they rose up from the surface of the plateaux directly in the path of the currents, the latter were compelled to flow along the side of the ridge opposed to them, and the result was a denudation of the plateau on one side of the ridge, while the accumulation of strata continued on the other side. We have here a possible explanation of the fact that we generally find an area of depression on one side of a mountain range—*i.e.*, that the latter forms or has formed at one time a coast range with a high plateau on the opposite side. On one side of the ridge we have a comparatively steep incline caused by the denudation of the plateau, on the opposite side the low and wide-spreading incline of the original plateau.

The continued action of the currents would ultimately result in the cutting through at right angles of the original plateaux, and in the formation of new plateaux following the direction of the meridian, while the ridges subsequently raised up on their surface would follow the same direction, stretching from north to south. The surface of our planet would now present the appearance of *primary* areas of elevation running parallel with the equator, with their ridges or mountain ranges stretching from east to west, backed up by plateaux on their polar or equatorial slopes, according as the denudation has been effected by equatorial or polar currents; and of *secondary* areas of elevation, following the direction of the meridian, with their mountain ranges running north and south, backed up by plateaux on their eastern or western slopes. In the case of the secondary areas of elevation, their direction along the meridian exposes them to denudation on both sides by equatorial and polar currents, hence the triangular shape of the present continents with their apex pointed towards the South Pole. Hence also the observed transfer of land from east to west. Both equatorial and polar currents are more powerful along the east coast than along the west coast of the continents, and the deposit of solid matter is in consequence least on the western side of an oceanic basin, greater on its eastern side, and greatest in its centre. This agrees with observed facts, for the western part of an oceanic basin is as a rule deeper than the eastern, while the plateaux are found in the centre.

Primary areas of elevation are exposed to denudation by equatorial currents upon their equatorial slopes, and by polar currents upon their polar slopes. The former currents being more powerful than the latter, the plateaux predominate upon the polar slopes, as we find it to be the case in the present continents; but the combined action of both equatorial and polar currents ultimately tends to break through the primary areas of elevation in the direction of the meridian, and to cut them up into separate continents. The latter would then present a combination of primary and secondary areas of elevation, with their respective watersheds and mountain ranges running at right angles to each other (Plate 4A).

FORMATION OF MOUNTAIN RANGES AND SUBMARINE RIDGES.—The application of the previous remarks to the configuration of the continents at present existing on the surface of the globe is too obvious to require further elucidation. There remains yet another question which the historian of our planet may be expected to answer, viz., the probable cause of the formation of ridges or mountain ranges, and of the creation of centres of volcanic activity.

Starting with Humboldt's and Sir Charles Lyell's definition of volcanic action as "the influence exerted by the heated interior of the earth on its external covering," we are led to inquire—What is the origin of this internal heat? The answer usually given is, that it proceeds from a primarily heated and fluid nucleus, to the gradual cooling of which we must attribute the formation of the solid external covering called the earth's crust. Observation has proved that the temperature of the earth-crust increases from the surface downwards, but the greatest depth at which it has been ascertained in mines and artesian wells does not exceed 360 fathoms (where it is found to remain constant at 75° F., or about 24° C.). On the other hand, the existence of a heated and fluid nucleus has been shown by recent calculations to be open to grave doubts, if not altogether impossible.

If in the absence of this cause of internal heat we proceed to look for another, we may possibly find it in an element which has been found invariably associated with volcanic action, and in a cause of heat the effects of which come under daily observation. This element is the *ocean*, and the cause of heat, *pressure*—namely, the pressure of superincumbent strata, both fluid and solid. Pressure, as an important factor in the structural development of the earth-crust, has not escaped the attention of the geologist, but the enormous pressure which the water contained in an oceanic basin must exert upon the bottom and the sides of the basin—a pressure roughly calculated to amount to one ton to the square inch for every mile of depth—has not been sufficiently insisted upon as an adequate cause of heat in the solid strata gradually accumulating at the bottom of the sea, and consequently as the primary cause of the various phenomena which are observed in connection with geological formations, such as stratification, cleavage, metamorphosis, and the final melting and eruption of strata in the form of fluid or semi-fluid matter.

We have seen that the deposits found at the bottom of the sea are different in their composition, according to the distance and the depth at which they are laid down. We may therefore expect that they will be affected differently by the heat developed under the pressure of the superincumbent ocean. If we attribute to pressure the observed increase of about 1° C. for every 20 fathoms in sub-aërial strata, we may expect a much greater rate of increase in strata subject to the enormous pressure of the ocean, and we may conceive the possibility of the existence of strata in a fluid or semi-fluid form at various depths below the bottom of the sea. The earth's crust would in that case be composed of strata of different degrees of solidity or fluidity, and the matter of the more fluid strata would, under the continuous influence of pressure, have a tendency to escape in a lateral direction. Now this lateral pressure will manifest itself at the point of least resistance—namely, upon the limit of an oceanic basin where the vertical pressure of the superincumbent ocean ceases altogether, or is sufficiently reduced to give way to the lateral pressure. The result will be an upheaval of the overlying strata along the margin of the oceanic basin or along the axis of a submarine plateau, and the formation of a mountain range or of a submarine ridge, both of which may or may not assume the character of an axis of volcanic eruption.

In this manner it may be explained why areas of elevation are older than the mountain ranges we find upon them, why mountain ranges are thrown up along sea-coasts—an almost certain evidence of the existence of lateral pressure exerted from the centre of the oceanic basin towards its margin—and also why the axis of a submarine plateau is generally found to coincide with an axis of volcanic eruption and a line of volcanic islands. In accordance with this view, we may conclude that where there are several ranges running parallel with the coast, the one nearest the coast will be of more recent origin than those further inland.

Several other phenomena, the explanation of which has until now been a matter of controversy, might be quoted in support of the oceanic origin of the dry land, but their discussion belongs more to the domain of geology than to that of physical geography. It is a significant fact that the results of recent microscopic examination of the materials which compose the different geological formations has led to a partial revival of a favourite theory of the early geologists—namely, the theory of the aqueous origin of rocks in opposition to the theory of their volcanic origin.

As the air of the atmosphere and the water of the ocean are distributed and renewed by a system of combined horizontal and vertical circulation, so the solid matter which composes the earth-crust is distributed and accumulated through the agency of oceanic currents, and also of atmospheric currents, but chiefly of the former, thus undergoing an unceasing process of disintegration and reformation. .To borrow an expression from the studio, water is the *vehicle* in which Nature dissolves her colours; while the ocean with his broad brush lays down the ground-tints of her pictures, the atmosphere with a more delicate pencil puts in the finishing touches.

www.ingramcontent.com/pod-product-compliance
Lightning Source LLC
Chambersburg PA
CBHW020926230426
43666CB00008B/1584